The History of the Big Ideas Mathematics

ESSENTIAL CALCULUS MATH PROBLEMS WITH APPLICATIONS |FAMOUS CONJECTURES IN MATHEMATICS

BASIC ARITHMETIC, GEOMETRY, ALGEBRA, PROBABILITY AND STATISTICS

MATHEMA

The History of the Big Ideas Mathematics

Contents

Introduction

Mathematics is a field that many people are avoiding, but there are some who had a passion for numbers and as well for making discoveries regarding equations, measurements, and other numerical solutions in history. They looked for ways to understand the world how it relates to numbers. Their contributions have been very important that time and beyond. Below is a list of names and accomplishments of the greatest mathematicians of all time.

Mathematics is an increasingly central part of our world and an immensely fascinating realm of thought.

Long before the development of the math that gave us computers, quantum mechanics, and GPS satellites, generations of brilliant minds — spanning from the ancient Greeks through the eighteenth century — built up the basic mathematical ideas and tools that sit at the foundation of our understanding of math and its relationship to the world.

On May 24, 2000, the Clay Mathematics Institute (CMI) presented seven major math problems. Each comes with a million dollar prize for whoever makes it through. Let's first find out what a conjecture means?

Larousse dictionary offers two definitions for the term conjecture :

- Assumption based on probability, but not checked by the facts; presumption, hypothesis: we are reduced to conjectures about our motivations.
- Hypothesis formulated on the accuracy or inaccuracy of a statement for which we don't have a proof yet.

Introduction

By definition, conjecture is the hypothesis formulated on the accuracy or inaccuracy of a statement which we do not yet know the demonstration. In other words, it is an "Open Question" for which a statement has been made. In mathematics, like other sciences, conjecture has always played a stimulating role. Each field of mathematics has its conjectures, more or less known more or less understandable.

The conjectures satisfies the following constraints:

- The statement is simple, understandable by the greatest number of mathematicians.
- Have resisted for a long time.
- Have generated new mathematics through the various attempts at solving.

In the first part, we talk about the major research themes of the 45 of the brightest mathematicians and philosophers in the world.

In the second chapter, we try to approach and simplify the statements of some of its conjectures chosen by great current mathematicians. These are the 7 problems considered to be the most important in mathematics today. One was solved by Russian mathematician Grigory Perelman in 2003.

PART 01

Famous mathematicians

Greek astronomer, geographer, philosopher and mathematician from the 3rd century BC, appointed at 30 to head the library of Alexandria, at the request of pharaoh Ptolemy III.

We owe him a measure of the circumference of the Earth quite valid, for his time, with an error of 1%. He is also the author of a method, called the Eratosthenes sieve, to determine the sequence of prime numbers.

He also created an astronomical catalog listing nearly 675 stars. All his observations of the sky allowed him to determine that the Earth was tilted. He is also known for having drawn one of the first maps of the world known in the third century BC.

ÉRATOSTHÈNE

~ 276 ~ 194 AV. J. – C.

His best known works relate to the circumference of the Earth. Eratosthenes is known to have been the first to give a reliable measure. Using certain geometrical properties of the alternate internal angles and based on the observation of the shadows cast at two different places, he managed to estimate it at 39,375 km, the current measurements give 40,075.02 km

Greek mathematician, surveyor, physicist and engineer. He was one of the main scientists and the greatest mathematician of Antiquity. In fact, he was one of the greatest of all time.

Archimedes' mathematical writings are little known in ancient times, but legends about it have been widely circulated through the ages. Several mathematicians who worked at the great library of Alexandria have read and quoted it repeatedly. Roman historians have reported many facts and gestures attributed to Archimedes, according to tradition or legend.

A large part of the subjects treated by Archimedes made an original contribution to the mathematics of his time. He is recognized for ten works dealing with various subjects such as balance in geometric figures, the squaring of the parabola, the sphere, the cylinder, etc.

We owe him works on the circle, precise approximations of π, studies on conics and in particular the parabola and a revolutionary approach to area and volume calculations.

ARCHIMÈDE DE SYRACUSE

~287– ~212 av. J.–C

The study of the circle where he determines a method of approximating π using regular polygons and shows the framing :

$$\frac{223}{71} \prec \pi \prec \frac{22}{7}$$

Archimedes' genius in mechanics and mathematics makes him an exceptional character from ancient Greece and explains the creation of legendary facts about him.

Mathematician member of the University of Alexandria from 125 to160 , author of Almageste, name of Arabic origin which means the largest. This astronomical work has acquired the same reputation as the Elements of Euclid in geometry.

The Ptolemy's Almagest has survived the ravages of time. In his astronomical and trigonometric works, Ptolemy divided the circle into 360° and the diameter into 120 parts, then each part into minutes, seconds and thirds, like the sexagesimal system of the Babylonians.

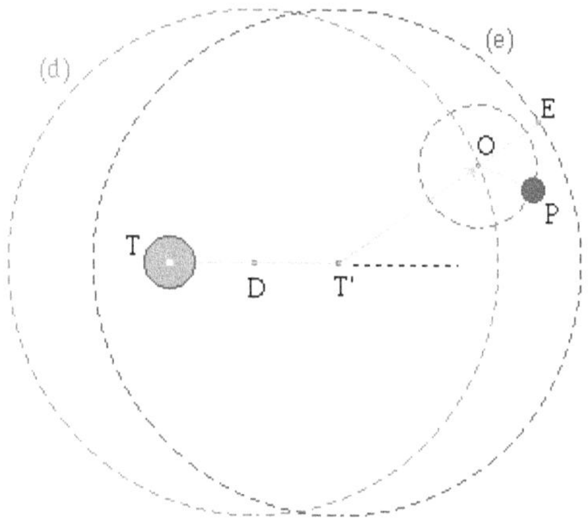

CLAUDIUS PTOLÉMÉE

$(\sim 85 - \sim 165)$

The theory of equants:

A planet P uniformly describes a small circle, its epicycle, around a center O describing itself around the Earth T, another circle (d) with center D, said to be deferent and with a larger radius R (main orbit of planet) so that if T 'is the symmetric of T with respect to D, equant point, and (e) the circle with center T' of radius R, then point E describes (e) uniformly. Under these conditions, it is the movement of O around D and T which is no longer uniform.

Christophorus Clavius is the Latin name for a German Jesuit scholar, mathematician and astronomer. He was the first to use parentheses to isolate mathematical expressions. At the request of Pope Gregory XIII, he was charged with drafting the foundations of a new calendar. This calendar was promulgated by Pope Gregory XIII in 1582 and gradually adopted around the world; it is still in use today under the name of the Gregorian calendar.

CHRISTOPHER CLAVIUS

(1538 – 1612)

Clavius proposed a proof that there can be no more than three dimensions in geometry, based on the fact that only three concurrent lines can be drawn from a point so that they are mutually perpendicular. He discovered and proved a theorem for a regular polygon with an odd number of sides which two centuries later enabled Carl Friedrich Gauss to construct a 17-sided polygon by ruler and compass.

Founder of the School of Mathematics at the University of Alexandria, after probably having received mathematical training at the Platonic Academy of Athens. He is recognized as the author of a fundamental work on geometry, numbers, elementary algebra, optics, etc. : Elements, one of the oldest known treatises on plane geometry and solids. This thirteen-book treatise systematically presents a set of theorems based on a system of axioms and postulates, as well as their demonstrations. Unfortunately, apart from the Elements, most of Euclid's works, translated into several languages, have been lost.

EUCLIDE

(~ 300 AV. J.C.)

Euclid cites five axioms in his work:

1st axiom: "By two points, it crosses a straight line"

2nd axiom: "Any segment can be extended as much as desired"

3rd axiom: "From any point, we can draw a circle of any radius"

4th axiom: "All right angles are equal"

5th axiom: "There is only one straight line parallel to d passing through A"

François Viète, a French mathematician, was the first mathematician to systematize the use of letters to denote unknown quantities. In his scriptures, vowels represented unknown quantities and consonants represented known quantities. Decades later, René Descartes, in his book La Géométrie (1637), used the first letters of the alphabet for known quantities and the last letters of the alphabet for unknown quantities. For example, the equation "$ax + by = c$", introduced by Descartes, is still used to describe the equation of a line.

FRANÇOIS VIÈTE

(1540–1603)

He ends with his formula giving π as an infinite product, where he believes to detect the proof that the Archimedes number cannot be constructed with a ruler and a compass. Starting from geometric considerations and by means of trigonometric calculations which he masters perfectly, he knows that:

$$\pi = 2 \times \frac{2}{\sqrt{2}} \times \frac{2}{\sqrt{2+\sqrt{2}}} \times \frac{2}{\sqrt{2+\sqrt{2+\sqrt{2}}}} \times \frac{2}{\sqrt{2+\sqrt{2+\sqrt{2+\sqrt{2}}}}} \times \cdots$$

Gabriel Cramer was a Swiss mathematician, professor of mathematics and philosophy at the Geneva Academy. Cramer's contributions to mathematics relate mainly to algebra and geometry, through his only published work, a treatise on curves entitled Introduction to the analysis of algebraic curved lines, published in Geneva in 1750. In this treatise finds in particular the method known today under the name of rule of Cramer for the resolution of the linear systems of equations, using what will be later called determinants.

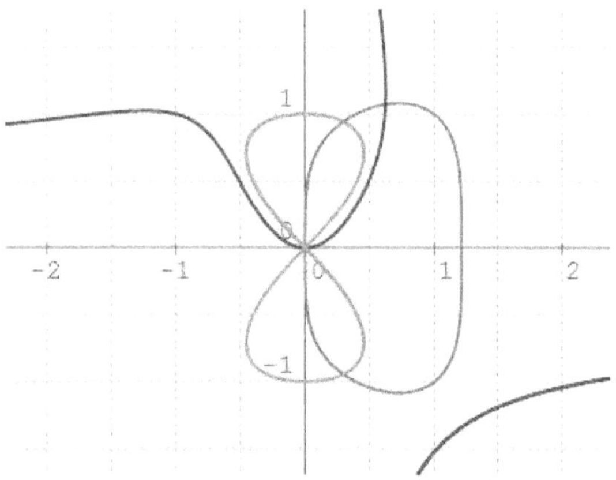

GABRIEL CRAMER

$(1704 - 1752)$

Cramer studies the algebraic curves defined in implicit form : by a polynomial equation in x and y of type $f(x, y) = 0$ by examining their singularities (multiple points, cusps, inflection points), their infinite branches, their asymptotes, their tangents.

Regarding double points, including cusps, Cramer shows this important result:

An algebraic curve of degree n has at most

$(n-1)(n-2) / 2$ double points or cusps.

German mathematician, to whom we owe set theory.

He based his theory on the principle of bijection between sets. He thus proved that the set of real numbers has a greater cardinality than the set of natural numbers. In other words, there are more real numbers than whole numbers.

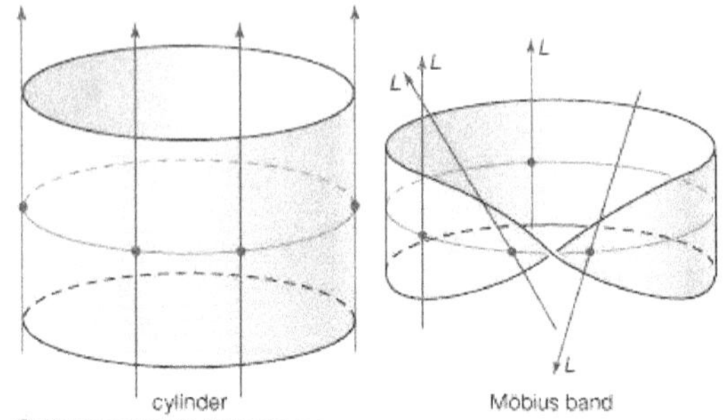

cylinder   Möbius band

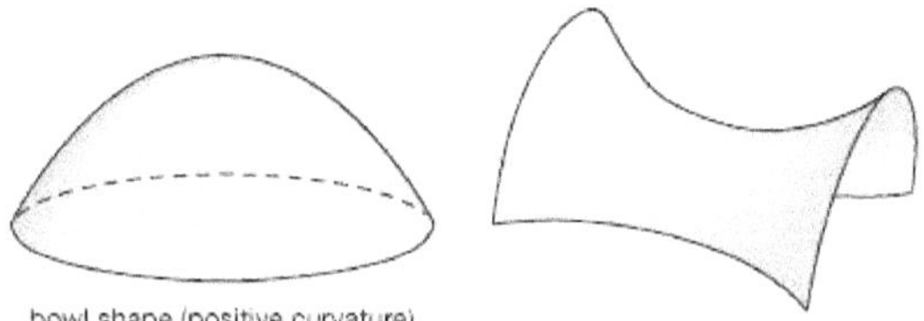

bowl shape (positive curvature)   saddle shape (negative curvature)

GEORG CANTOR

(1845–1918)

The early 20th century saw the emergence of a number of theories whose power and utility reside in large part in their generality. Typically, they are marked by an attention to the set or space of all examples of a particular kind. (Functional analysis is such an endeavour.) One of the most energetic of these general theories was that of algebraic topology. In this subject a variety of ways are developed for replacing a space by a group and a map between spaces by a map between groups.

15th century German mathematician who published a commercial arithmetic at a very young age entitled Behende und hüpsche Rechenung auff allen Kauffmansschafft. This is said to be the oldest known printed book in which our familiar symbols for the addition

"+ " and the subtraction "- " appear for the first time to indicate an excess or deficiency in warehouse measurements.

Johann Widmann taught arithmetic and algebra at the University of Leipzig. In a treaty of commercial arithmetic published in Leipzig (1489), he uses the signs + and -: to express credit balances (gains) and debit balances (debts). These ratings will be renewed by Riese a few years later to designate addition and subtraction.

JOHANNES
WIDMANN

(1460 – 1498)

It will however be necessary to wait for the French Viète, half a century later, for a systematic use of this symbolism that did not use the great Italian algebraists like Cardan or Bombelli still preferring the p and the m accented in the manner of the tilda (~)

Greek mathematician and philosopher from Alexandria. Daughter of Théon of Alexandria, commentator on the Elements of Euclid and last director of the library of Alexandria, Hypathy was the first known mathematician in the history of mathematics, recognized in his time, in addition to shining in medicine and in philosophy. She is said to have written comments on Diophantine Arithmetic and the Appolonius Conic Sections.

Daughter of the mathematician Théon of Alexandria, a famous professor at the Alexandria Museum, she was initiated by her father to mathematics and philosophy and studied in Athens, under the tutelage of Asclepigenia, master of philosophy. She then returned to Alexandria and directed, around the year 400, the Neoplatonic School of Alexandria, where she taught the precepts of Plato and Aristotle to disciples from the wealthy and cultivated strata of Alexandrian society.

HYPATIE
D'ALEXANDRIE

(~ 355 – 415)

She will remain a virgin all her life. His way of life has traits that resemble the immodesty of cynics. She walks the streets of Alexandria offering her philosophical teaching and extinguishes the too passionate love of a disciple by showing him the stained linens of his menstrual flow.

Swiss mathematician. In his treatise on algebra, in 1659 , he introduced the symbol * as a symbol of multiplication and the symbol ÷ as a symbol of division $(a \div b)$.

According to D. M. Burton, Pell introduced the notation $(a \div b)$ in England (1668). The notation was adopted by Wallis.

In the $(a \div b)$ notation, a is the dividend and b is the divisor. The result is the quotient.

In Germany, the usual fractional notation a / b , Leibniz used (1684) the colon, as in $10: 2 = 5.$

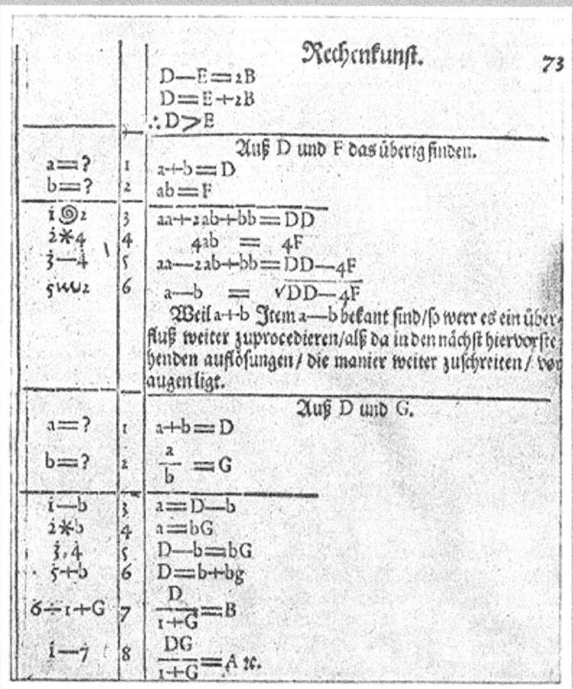

Rechenkunst. 73

		D—E = 2B D = E + 2B ∴ D > E
		Auß D und F das übrig finden.
a = ?	1	a + b = D
b = ?	2	ab = F
1 ⊚ 2	3	aa + 2ab + bb = DD
2 * 4	4	4ab = 4F
3 — 4	5	aa — 2ab + bb = DD — 4F
5 ωω 2	6	a — b = √DD — 4F
		Weil a + b Item a — b bekant sind/so werr es ein überfluß weiter zuprocedieren/alß da in den nächst hiervorstehenden auflösungen/die manier weiter zuschreiten/vor augen ligt.
		Auß D und G.
a = ?	1	a + b = D
b = ?	2	$\frac{a}{b}$ = G
1 — b	3	a = D — b
2 * b	4	a = bG
3, 4	5	D — b = bG
5 + b	6	D = b + bg
6 ÷ 1 + G	7	$\frac{D}{1+G}$ = B
1 — 7	8	$\frac{DG}{1+G}$ = A 2c.

JOHANN RAHN

(1622–1676)

LA DIVISION

Calculons la valeur d'une part ou d'un objet.

138. — *Problème.* — ***Une maman partage également 15 oranges entre ses 3 enfants. Combien en donne-t-elle à chacun?***

Fig. 62. — La division.

La maman peut donner une orange à chaque enfant, ce qui fera 3 oranges distribuées; puis encore une orange à chacun, ce qui fera 2 fois 3 oranges distribuées... et ainsi de suite jusqu'à ce qu'il n'y ait plus d'oranges. Chaque enfant recevra donc autant d'oranges que 15 oranges contiennent de fois 3 oranges. On peut faire rapidement ce *partage* ou *division* grâce à la table de multiplication.

On dit : En 15 combien de fois 3? 5 fois (puisque 5 fois 3 font 15).
On écrit : 15 : 3 = 5
On prononce: 15 divisé par 3 égale 5
15 est le *dividende*; 3 est le *diviseur*; 5 est le *quotient*.

139. — ***Faire une division, c'est chercher combien de fois un nombre appelé dividende en contient un autre appelé diviseur. Le résultat s'appelle quotient.***

la division expliquée à l'école primaire par MM. Royer et Court
(*Arithmétique*, Ed. A. Colin, 1931)

17th century English mathematician. His work was a precursor to that of Newton.

He is also a pioneer in phonetics, deaf education and speech therapy. We owe him the symbol of infinity (∞) as it is used today. He was one of the founders of the famous Royal Society (Oxford).

His work mainly concerns differential and integral calculus where he introduces the general Wallis integrals: $\int_0^{\frac{\pi}{2}} \sin^n(x)\mathrm{d}x$

He solved the problem of the quadrangle vault (1692), posed by Vincenzo Viviani: find a window in a hemispherical vault so that the rest of the vault is quadrangle, that is to say whose area can be written C^2, where C is a constructible number with ruler and compass.

JOHN WALLIS

(1616–1703)

The Wallis product formula

$$\frac{\pi}{2}=\prod_{n=1}^{\infty}\frac{4n^2}{4n^2-1}$$

is equivalent to the generalized continuous fraction development of $\frac{4}{\pi}$ found by William Brouncker.

13th century Italian mathematician.

Leonard de Pisa, better known as Fibonacci (son of Bonaccio), often accompanied his father, an Italian merchant, on his travels to Syria, Greece and Egypt, in particular. During his travels, he became convinced that the Hindu calculation methods were by far the best and most effective. Back home in 1202 , he composed his still famous work, the Liber abaci, a bit confusing title, because this Book of the abacus deals mainly with algebraic methods and problems in which the interest of the use of Indo figures -arabes is highlighted. Divided into fifteen chapters, the Liber abaci emphasizes the nine Hindu symbols of numeration as well as the zero sign, whose name comes from zephirum, Latin form of the Arabic sifr.

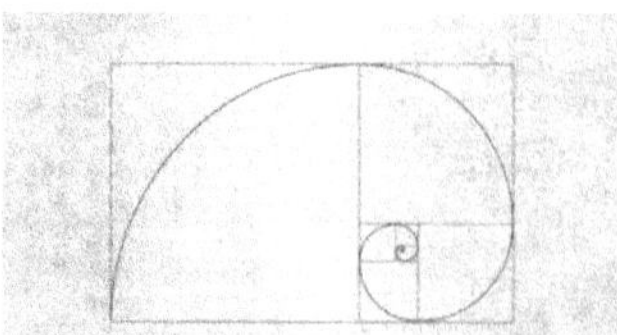

LEONARDO
FIBONACCI

(~ 1180 – 1250)

In 1220, Fibonacci wrote a new work, Practica geometricae. He presented new theorems, commented on eight books of Euclid's Elements, exposed a demonstration of Pythagoras' theorem and studied the duplication of the cube. There is also the remarkable formula of Heron which gives the area of a triangle as a function of the length of its sides (a, b and c) and the half perimeter $\left[p = (a + b + c) \div 2\right]$:

$$A = \sqrt{p(p-a)(p-b)(p-c)}$$

German mathematician who lived in the 19th century. It is at the origin of the use of the two vertical bars to write the absolute value of a mathematical expression like

$|21|$ or $|x+2|$, etc.

On July 18, 1872, the mathematician Karl Weierstrass presented before a dumbfounded crowd not one, a whole family of continuous functions derivable nowhere! There were two others then, but were not published (that of Bolzano and that of Cellérier, which is a special case of that of Weierstrass).

The Weierstrass curves are given by the following formula:

$$W(x)=\sum_{k=0}^{\infty} a^k \cos\left(b^k \pi x\right)$$

$$0 \prec a \prec 1;\ b\ odd\ ;\ ab \succ 1$$

KARL WEIERSTRASS

(1815–1897)

for example $a=\frac{1}{3}$ and $b=5$. We then have:

$$W(x)=cos(\pi x)+\frac{cos(5\pi x)}{3}+\frac{cos(25\pi x)}{9}+\ldots$$

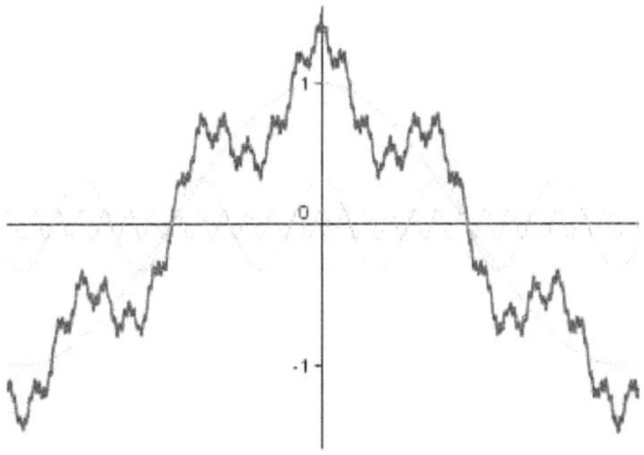

Fonction de Weierstrass

En gris les sinusoïdes à sommer pour obtenir la fonction

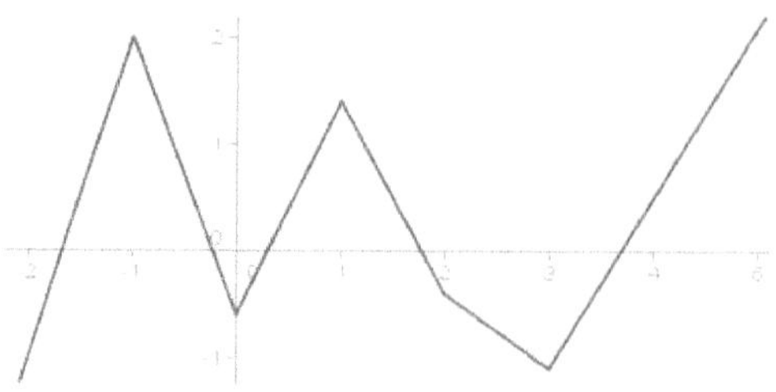

Graphe d'une fonction zigzag

Fonction continue mais non dérivable en -1, 0, 1, 2 et 3

Mathematician and scientist who worked in the library of Caliph al-Mâmûn in Baghdad.

As was often the custom in the Persian and Arab populations, the patronym came from the place of birth of the individual. Muhammad, son of Mūsā, is from Khuwārizm (today the city of Khiva, Uzbekistan) which gave him his surname of al-Khuwārizmī, which means from Khuwāriz.

His works have not reached us in their original versions, but we know translations in Latin. One of the most interesting is certainly the Liber algoritmi de numero Indorum which presents the numerical calculation rules modeled on Hindu algorithms, which used a positional numeration system based on ten (using ten symbols). This new notation gave rise to the word "algorithm". However, the main work of al-Khuwārizmī is Hisâb al-jabr wa'l-muqqâbala whose title means "science of transposition and reduction", where the term "al-jabr" has become "algebra".

MUHAMMAD IBN MŪSĀ AL-KHUWĀRIZMĪ

$(\sim 780 - 850)$

In order to justify the validity of his algebraic calculations, Al-Khawarizmi proved beforehand, as Euclid did before, certain usual remarkable identities using the geometric support of areas. Indeed, we easily prove by this means the two formulas :

$$(a + b)^2 = a^2 + 2\,ab + b^2,$$

$$(a - b)^2 = a^2 - 2\,ab + b^2$$

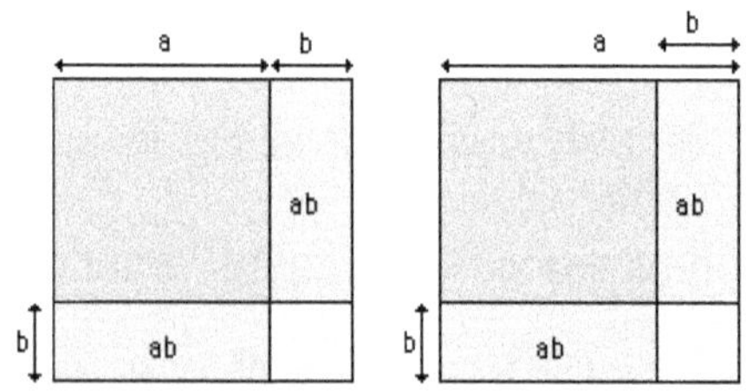

Swiss mathematician and physicist recognized in particular for his work on infinitesimal calculus and graph theory. He is responsible for much of the terminology and notation used in modern mathematics, particularly in the area of functions.

An early student, Euler published his first dissertation at the age of 18. Without more experience, he tackles complex problems to which he brings original solutions. At 20, he joined the St. Petersburg Academy where he held a position as professor of physiology. At 26, he occupied the chair of mathematics left vacant after the death of his friend Daniel Bernouilli. Subsequently, his name is found in almost all branches of mathematics and other scientific fields for which he has published nearly 900 works. He was a virtuoso mathematician with an out of the ordinary memory capable of performing mentally complex calculations and of giving complicated formulas or arithmetic calculation results from memory.

He was the most prolific mathematician of his time; his complete work includes 886 books and articles.

LEONHARD EULER

$(1707-1783)$

We owe him the notation $f(x)$ (1734), e for the base of natural logarithms (1727), i for the square root of -1 (1777), π for pi, Σ as a summation symbol (1755) etc. He also introduced beta and gamma functions, integrating factors for differential equations etc.

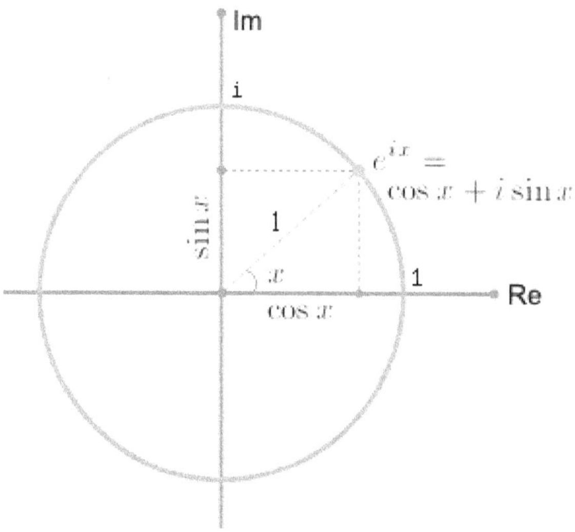

Norwegian mathematician.

We owe to Abel the notion of an algebraic number, solution of a polynomial equation with rational coefficients. Abel also gave its name to the groups whose internal composition law is commutative.

Still a student, he demonstrated the impossibility of solving by radicals the general algebraic equation of the 5th degree. He rigorously studied the convergence domain of the binomial series and transformed the theory of elliptic integrals by studying their inverse functions and considering complex values of the variable. He extended his research to abelian integrals.

NIELS HENRIK ABEL

(1802–1829)

Elliptical integrals. The elliptical integrals, are the integrals of the form: $\int R\left(x, \sqrt{P(x)}\right) dx$

For R a rational fraction and P a polynomial of degree 3 or 4 .

He notes that the length of BERNOULLI's lemniscate leads to the integral: $\int_0^1 \frac{1}{\sqrt{1-x^4}} dx$

It then makes the link with the formula:

$$\text{Arcsin}(t) = \int_0^t \frac{1}{\sqrt{1-x^2}} dx$$

By analogy with the restriction of the sine function to $[-\pi/2;\ \pi/2]$, reciprocal of Arcsin, he then constructs a reciprocal periodic function of the indefinite integral studied.

French mathematician and astronomer, of Basque origin, born around 1580 and died in Paris between 1643 and 1644.

He introduced the symbol of orthogonality ($\perp$) and introduced the term parallelepiped.

Hérigone may have been the first to introduce the mathematical symbol to express an angle. He used both the symbol below and recorded the use of "<" as a symbol denoting "less than." Hérigone also created a number alphabet for remembering long numbers in which phonemes were assigned to different numbers, while the vowels were supplied by the memorizer: (Article Herigone's mnemonic system).

$1\,(t,\,d),\ 2\,(n),\ 3\,(m),\ 4\,(r),\ 5\,(l),\ 6\,(j,\,ch,\,sh),$
$7\,(c,\,k,\,g),\ 8\,(f,\,v,\,ph),\ 9\,(p,\,b),\ 10\,(z,\,s)$

PIERRE HÉRIGONE

(1580 – 1644)

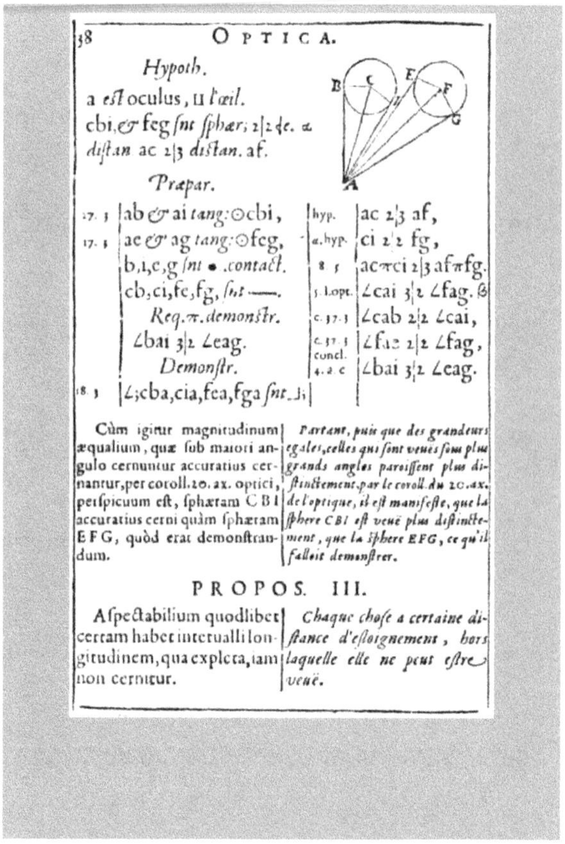

38 OPTICA.

Hypoth.

a *est* oculus, ⊔ *l'œil.*

cbi, & feg *snt sphær;* 2|2 *&c.* & *distan* ac 2|3 *distan.* af.

Præpar.

17. 3 | ab & ai *tang:* ⊙cbi,
17. 3 | ae & ag *tang:* ⊙feg,
| b,i,e,g *snt* • *.contact.*
| cb,ci,fe,fg, *snt* —.
| *Req.π.demonstr.*
| ∠bai 3|2 ∠eag.
| *Demonstr.*
18. 3 | ∠;cba,cia,fea,fga *snt* ⌞;

hyp. | ac 2|3 af,
α.hyp. | ci 2|2 fg,
8. 5 | ac π ci 2|3 af π fg.
5. l.opt. | ∠cai 3|2 ∠fag.
c. 37. 3 | ∠cab 2|2 ∠cai,
c. 37. 3 | ∠fae 2|2 ∠fag,
concl. 4. 2. c | ∠bai 3|2 ∠eag.

Cùm igitur magnitudinum æqualium, quæ sub maiori angulo cernuntur accuratius cernantur, per coroll. 20. ax. optici, perspicuum est, sphæram CBI accuratius cerni quàm sphæram EFG, quòd erat demonstrandum.

Partant, puis que des grandeurs egales, celles qui sont veuës sous plus grands angles paroissent plus distinctement, par le coroll. du 20. ax. de l'optique, il est manifeste, que la sphere CBI est veuë plus distinctement, que la sphere EFG, ce qu'il falloit demonstrer.

PROPOS. III.

Aspectabilium quodlibet certam habet interualli longitudinem, qua expleta, iam non cernitur.

Chaque chose a certaine distance d'esloignement, hors laquelle elle ne peut estre veuë.

Greek mathematician who lived in the 6th century BC. Born in the first half of the sixth century BC, on the Aegean island of Samos in Ionia, very close to Miletus. Died around 500 BC in Metapontum where he went into exile. The dates are approximate because Pythagoras, although repeatedly cited by the ancients, did not write anything of himself and there is no biography of him written in his time.

Pythagoras relationship :

Relationship between the measurements $a, b, and\ c$ of the sides of a right triangle:

$a^2 = b^2 + c^2$, where a represents the measurement of the hypotenuse and b and c the measurements of the two sides of the right angle.

With a modern look, one would be inclined to say that Pythagoras (Πυθ-αγόρας, "announced by the Pythia") was in his time a guru, master of thought, model, at the same time an accomplished Olympian athlete, first philosopher, pedagogue, mathematician, musician, both worshiped by the disciples of his school and hated by the politicians who chased him and ordered the destruction of all the buildings that housed his school.

PYTHAGORE

~580– ~495 AV. J.–C

Guru also because of the secret and sectarian character of the Pythagorean school where everything was pooled, where knowledge was attributed to the master, where as in an initiation organization, the members must go through various hierarchical degrees applying, neophyte, acousmatician (listener), then mathematician (scientist).

The symbol of recognition of the followers of the Pythagorean school was the pentagram, or regular star pentagon.

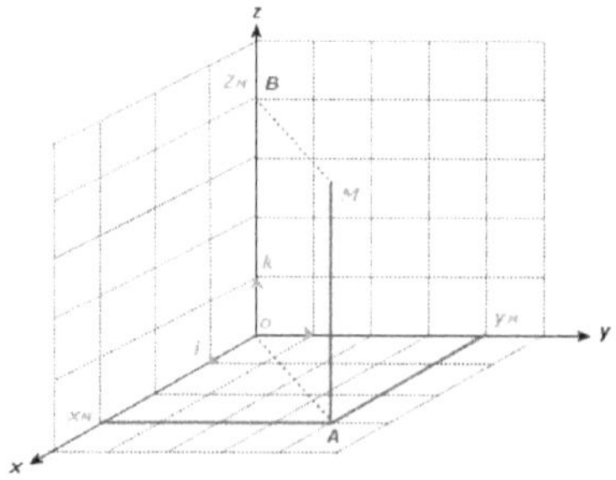

16th century Italian mathematician. Raphaël Bombelli is a son of a Bologna merchant and becomes an engineer (he dries up swamps in particular). He is employed by a Roman, Alessandro Ruffini, to carry out a long work which will experience an interruption of a few years, which gives him time to write an algebra from the 1560*s*. He was the first to use hooks in his work " Algebra ". He was one of the great Italian algebraists of the 16th century. He used the four basic operations with complex numbers and he used continuous fractions to calculate square roots and cube roots.

The first three books of Bombelli's Algebra were published in 1572 , a few months before his death. Bombelli was therefore the first to spread the problems of Diophantus in the West, although these, scattered throughout the Algebra and much more symbolized, will be identified later. Above all, it favored the emergence of a more abstract and theoretical formulation of algebra.

RAPHAEL BOMBELLI

(1526 – 1572)

L'ALGEBRA
OPERA
Di RAFAEL BOMBELLI da Bologna
Diuisa in tre Libri.
Con la quale ciascuno da se potrà venire in perfetta cognitione della teorica dell'Arimetica.
Con vna Tauola copiosa delle materie, che in essa si contengono.
Posta hora in luce à beneficio delli Studiosi di detta professione.
IN BOLOGNA,
Per Giouanni Rossi. MDLXXIX.
Con licenza de' Superiori.

The solution of the irreducible case of the cubic equation comes back to him. 4th degree equations are also treated by FERRARI methods.

William Jones was a Welsh mathematician. He has the distinction of having taught mathematics on a boat between 1695 and 1702. After these trips, he became a professor of mathematics in London. The symbol π (pi), designating the relationship between the circumference of a circle and its diameter, was first used by William Oughtred in 1647 and it first appeared in 1706 in the book " Synopsis Palmariorium Matheseos " written by William Jones. In this book there was also a list of the first 100 decimals of π calculated by John Machin $(1680-1751)$. Jones is believed to have used the Greek letter π because it was the first letter of the Greek word perimetron (περίμετρος] which designated the measurement of the perimeter. This symbol was definitively imposed by Euler in 1748 .

WILLIAM JONES

(1675 – 1749)

It appears in a book by William Jones *in* 1706

in the circle, the diameter is at the circumference as 1 to

$$\left(\frac{16}{5}-\frac{4}{239}\right)-\frac{1}{3}\left(\frac{16}{5^3}-\frac{4}{239^3}\right)\ldots=3{,}14159=\pi$$

In modern writing:

$$\lim_{n\to\infty}\sum_{k=0}^{n}\frac{(-1)^k}{2k+1}\left(\frac{16}{5^{2k+1}}-\frac{4}{239^{2k+1}}\right)=\pi$$

There are various other ways of finding the *Lengths*, or *Areas* of particular *Curve Lines*, or *Planes*, which may very much facilitate the Practice; as for Inftance, in the *Circle*, the Diameter is to Circumference as 1 to $\overline{\frac{16}{5}-\frac{4}{239}}-\frac{1}{3}\overline{\frac{16}{5^3}-\frac{4}{239^3}}+\frac{1}{5}\overline{\frac{16}{5^5}-\frac{4}{239^5}}-$, &c. = 3.14159, &c. $=\pi$. This *Series* (among others for the fame purpofe, and drawn from the fame Principle) I receiv'd from the Excellent Analyft, and my much Efteem'd Friend Mr. *John Machin*; and by means thereof, *Van Ceulen*'s Number, or that in Art. 64.38. may be Examin'd with all defireable Eafe and Difpatch.

French philosopher, mathematician, biologist and physicist.

First known for his Discourse on the Method of the Philosophy of Science, René Descartes influenced many other scientific fields, despite the religious censorship of his time. Thus, he replaced in his publications the syllogism inherited from the Aristotelian era and which had prevailed throughout the Middle Ages by the mathematical method. He thus proposes to extend the certainty of mathematical reasoning to the whole of knowledge, by affirming that all phenomena must be able to be explained by mathematical reasons, that is to say by figures and movements in accordance with "Laws". Hence the expression "having a Cartesian spirit".

The notion of location in a plane, called the Cartesian plane, is proposed in the first book of the Geometry of Descartes, published in 1637 . Also, its new notation to describe the equations and algebraic operations made it possible to kick off of the development of modern mathematics.

RENÉ DESCARTES

(1596 – 1650)

Descartes was also the first to use an exponent to designate a power, such as 8^4, in his book "La Géométrie", published in 1637. .

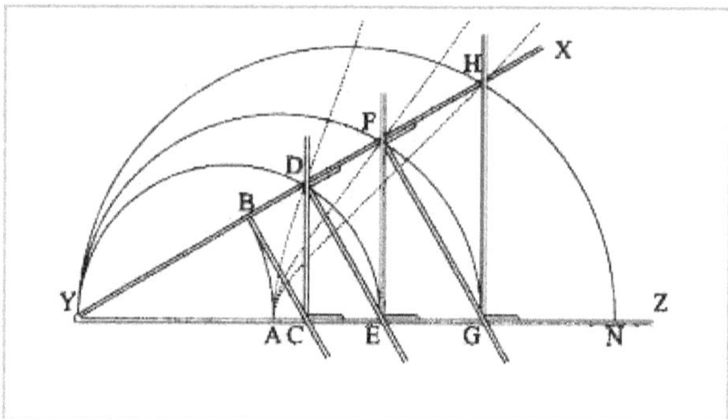

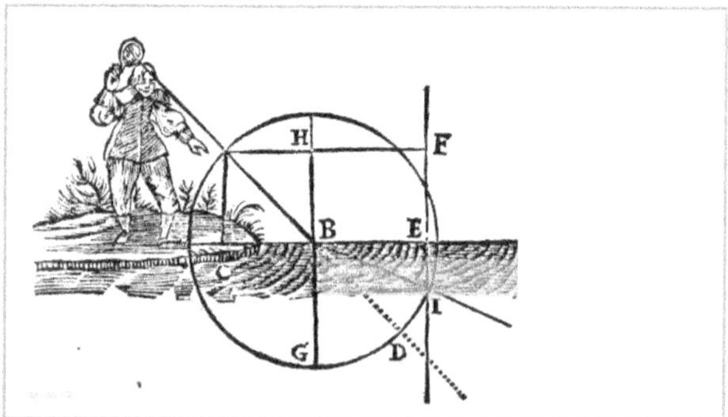

16th-century Welsh mathematician, trained at the University of Cambridge and the University of Oxford.

In a work published in 1557 , he introduced algebra, invented the equality symbol "=" and introduced the addition symbol "+" invented before him by the mathematician Johann Widmann.

Recorde's writings were all in the vernacular and assumed the form of a dialogue between teacher and pupil. He appears to have envisaged a publishing enterprise that would introduce each branch of mathematics.His first publication was The Ground of Artes (1543, enlarged 2nd ed.1552), an effective treatise on elementary arithmetic, with a section for merchants on using the abacus. In 1551 there appeared Recorde's textbook on elementary geometry, Pathwaie to Knowledge, a reworking of the first four books of Euclid's Elements.

Howbeit, for easie alteratiō of *equations*. I will propounde a fewe exāples, bicause the extraction of their rootes, maie the more aptly bee wroughte. And to auoide the tediouse repetition of these woordes : is equalle to : I will sette as I doe often in woorke use, a paire of paralleles, or Gemowe lines of one lengthe, thus: ======, bicause noe. 2. thynges, can be moare equalle. And now marke these nombers.

14.ꝛ.—+—.15.ϑ======71.ϑ.

ROBERT RECORDE

(1512–1558)

The Gate of Knowledge, which dealt with measurement and use of the quadrant, is known only through references in later books. Then followed The Castle of Knowledge (1556), a treatise on the sphere and Ptolemaic astronomy—though it also made favourable mention of the heliocentric theory of Nicolaus Copernicus (1473–1543), promising to deal with the subject at greater length in a subsequent work. His last work, The Whetstone of Witte (1557), was an advanced treatise on arithmetic as well as an introduction to algebra and used his new symbol for equality $(=)$.

Auguste Ferdinand Möbius was a mathematician and theoretical astronomer at the University of Leipzig in Germany.

In 1809 he studied mathematics and astronomy, successively at the universities of Leipzig, Göttingen (there was Carl Friedrich Gauss as professor) and Halle.

He is mainly known for his discovery of the Möbius strip, a non-orientable two-dimensional surface with only one face when it is immersed in a three-dimensional Euclidean space.

It was discovered independently by Johann Benedict Listing around the same time.

Möbius inversion function and formula :

The function of Möbius μ is the function defined on the integers greater than or equal to 1 by :

AUGUST
FERDINAND
MÖBIUS

(1790 – 1868)

(•) $\mu(1) = 1$

(•) if n has a square factor, then $\mu(n) = 0$

(•) if n is the product of all distinct prime numbers, $n = p_1 \cdots p_k$, then $\mu(n) = (-1)^k$

Thus, $\mu(2) = -1$, $\mu(4) = 0$, $\mu(6) = 1$.

The μ function checks the following property:

$$\forall n > 1, \sum_{d \backslash n} \mu(d) = 0$$

Möbius inversion formula: Let $f : \mathbb{N}^* \to C$ and set, for $n \geq 1$, $g(n) = \sum_{d \backslash n} f(d)$.

So : $f(d) = \sum_{d \backslash n} \mu\left(\frac{n}{d}\right) g(d)$.

William Oughtred, an English mathematician and theologian, vicar of Shalford and rector of Albury, was the first to use the symbol $\angle\angle$ to designate an angle in his book Trigonometria (1657). This symbol replaced the $<$ which was previously used and which had been recommended by Pierre Herigone, a French mathematician, in his book " Cursus Mathematicus " (1634).

His work has mainly focused on the applications of algebra to geometry, on the construction of equations, on the formation of powers.

He is the author of the abbreviated multiplication process well known as the Oughtred rule. He introduced the cross (x) to signify the multiplication operation. He also invented circular slide rules.

We call the Oughtred Rule a process of abbreviated multiplication which makes it possible to obtain, to a unit near a certain order, the product of two whole or decimal numbers, and which has become classic. We reproduce it here, according to Serret, in a form which answers almost all the cases which arise in practice, and which moreover can be easily modified:

WILLIAM OUGHTRED

(1574 – 1660)

For example, multiply 31.415926535897 by 986.96070733, the product to be obtained to the nearest 0.001. The operation will be arranged as follows:

```
  31415926535897
33707069689
----------------
 2827433385
  251327408
   18849552
    2827431
     188490
       2198
         21
----------------
3100628485
```

The sought product is 31006,285 to the nearest 0.001.

French mathematician, physicist and philosopher.

She was in contact with other mathematicians of her time such as Gauss and Fourrier, after having become familiar with the works of Euler, Newton and Fermat. To gain acceptance in the scientific society of her time, she had to hide her identity by borrowing the name of Antoine Auguste Le Blanc, a former student of the École Polytechnique. The field of mathematics owes him the theorem which bears his name - Sophie Germain's theorem - which sets out a sufficient condition relating to a prime number p so that, if three integers x, y and z form a solution of the equation $x^p + y^p = z^p$, then at least one of the three numbers x, y and z is divisible by the square of p. This statement will reduce the number of solutions to Fermat's last theorem. We will discover later that she had approached very closely the solution of this famous theorem.

SOPHIE GERMAIN

(1776–1831)

Sophie Germain Identity :

Patrick Corn, Anandmay Patel, Worranat Pakornrat, and 1 other contributed

The following algebraic identity is due to Sophie Germain:

$$\begin{aligned}x^4+4y^4 &= \left(x^2\right)^2+\left(2y^2\right)^2+4x^2y^2-4x^2y^2\\ &= \left(x^2+2y^2\right)^2-\left(2xy\right)^2\\ &= \left(x^2+2y^2+2xy\right)\left(x^2+2y^2-2xy\right)\\ &= \left((x+y)^2+y^2\right)\left((x-y)^2+y^2\right)\end{aligned}$$

She discovered this identity in her explorations related to Fermat's last theorem and primality testing. The identity is often used to find factors of integers of a certain form, and is common in contest mathematics.

African American physicist, mathematician and space engineer.

She has contributed to the aeronautical and space programs of the National Aeronautics and Space Administration $(NASA)$. Renowned for its reliability in computerized astronomical navigation, it has conducted technical work at $NASA$ which has spanned decades. In 2017 , she was still alive and was 99 years old. A $NASA$ building was named after him. During this period, she calculated the trajectories, launch windows and emergency plans for numerous Mercury flights, including the first missions of John Glenn and Alan Shepard, and the Apollo 11 flight to the Moon in 1969 until to the space shuttle program. His calculations were essential to the effective conduct of these missions. Johnson also calculated plans for a mission to Mars.

In 2015 Johnson was awarded the Presidential Medal of Freedom. The following year, she was included in the BBC's 100 Women of Exception series.

KATHERINE JOHNSON

(1918–2020)

Through his calculations, Man walked on the Moon

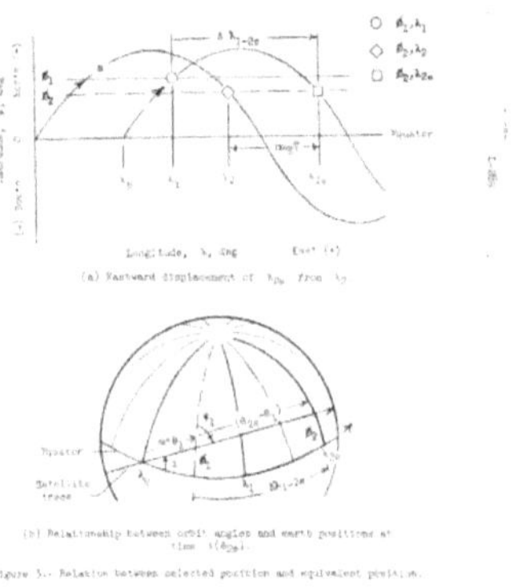

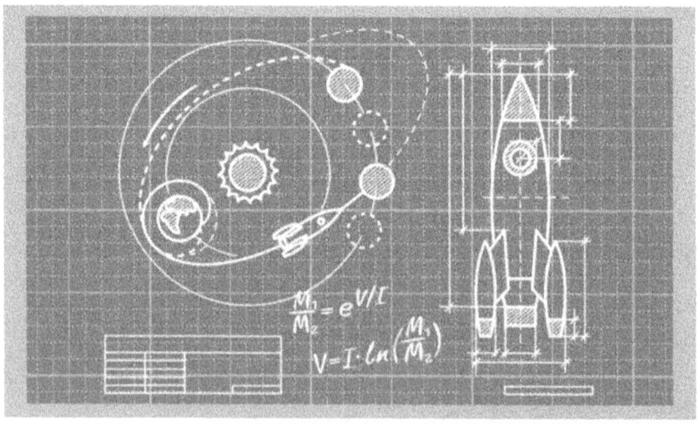

Mathematician and computer scientist.

Ada Lovelace became interested in mathematics at a very young age, unlike the habits of young nobility girls. After a meeting with mathematician Charles Babbage who was interested in machines capables of performing complex calculations quickly, she was fascinated by these machines and saw various possible developments. In 1842 , she was asked to translate into English a thesis on Babbage's analytical machine, thesis proposed by the Italian mathematician Federico Luigi. Invited by Babbage to add her personal notes to this translation, she introduces in these additions the concept of computer program to calculate Bernouilli numbers. This program is still considered today as the very first computer program in the world.

ADA LOVELACE

(1815–1852)

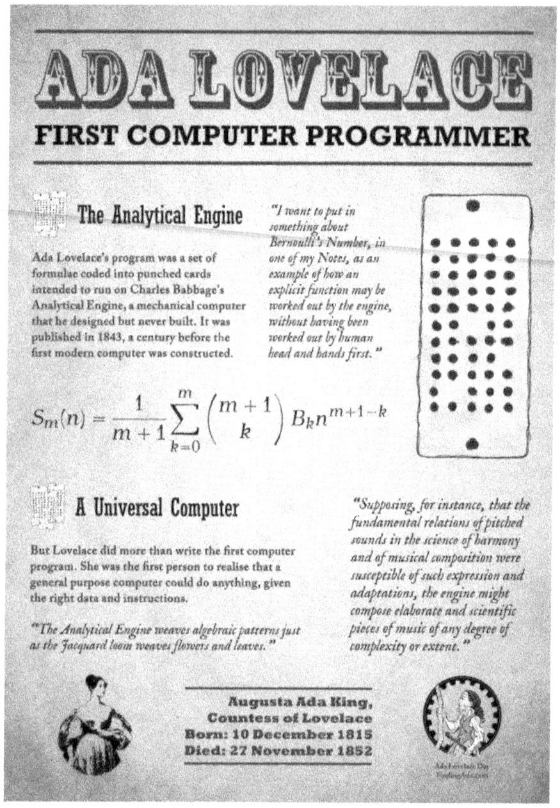

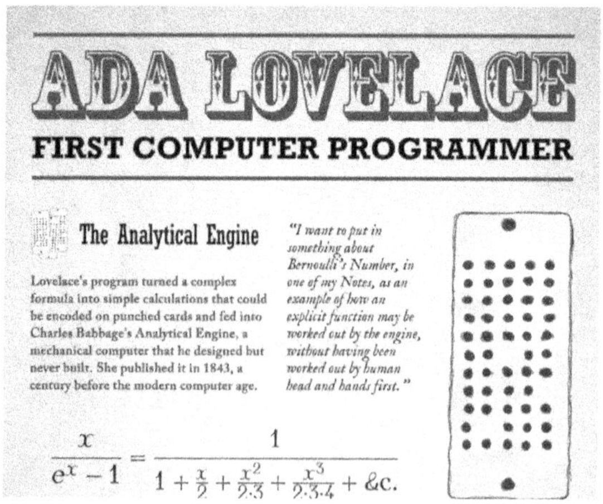

German mathematician and physicist who revolutionized the theories of rings, bodies and algebras.

From 1908 to 1919, after completing her mathematics thesis, she worked as a volunteer at the Institute of Mathematics in Erlangen, Bavaria, during which time she became interested in the theory of algebraic invariants and number bodies. Later admitted as a lecturer at the same university, she began work that would change the face of algebra and become essential in the movement of modern mathematics. In 1933, when she was universally recognized for her knowledge of algebra and yet banned from university work by the Nazi regime, she emigrated to Pennsylvania, to the United States. During this last period of her life, she will publish the results of her research on non-commutative algebras and hypercomplex numbers.

While in the United States, Emmy had presented readings at the Princeton Institute for Graduate Studies. She died in 1935 at the age of 53 from complications from surgery for an ovarian cyst. In a letter to the New York Time, Albert Einstein describes her as "the most significant creative genius that we have known since women gained access to higher education." Amalie Emmy Noether is recognized as one of the greatest mathematicians of the $20th$ century.

AMALIE EMMY NOETHER

(1882 – 1935)

In the 1920*s* , Emmy Noether deepened his research in abstract algebra and published important articles on ideals (in mathematics, an ideal is a remarkable subset of a ring). Brilliant and noticed, his works and his publications earned him many praises from fellow mathematicians. Irving Kaplansky, in particular, called his article of 1921 , which gave birth to the term Noetherian ring, "revolutionary". Norbert Wiener considers Emmy to be "the greatest mathematician who has ever lived, and the greatest living female scientist in all fields, and a scholar of the same standard, at least, as Madame Curie".

Aryabhata was a famous astronomer and also the first great Indian mathematician. It is known to us by an important treaty, translated in Europe in the 19*th* century, called the Aryabhatiya, written in Sanskrit (sacred language of the Brahmans) in 499 and relating to astronomy and mathematics.

Contrary to the geocentric doctrine of Ptolemy, then widespread and inherited from Aristotle, that the Earth is motionless at the center of the universe, Aryabhata affirmed the rotation of the Earth. In his Aryabhatiya, he describes the algorithms for extracting square roots and cube roots; it makes use of a positional decimal system whose graphics are close to ours and where the use of zero appears implicitly. This short treatise on astronomy, published in 499 , contains a summary of Indian mathematics and in particular 66 theorems of arithmetic, algebra, plane and spherical trigonometry. Results on continuous fractions, quadratic equations and power series are also presented.

To do this, Aryabhata invents a system of representation of numbers based on the consonants of the Indian alphabet, possibly followed by vowels for large numbers.

ARYABHATA

(476 – 550)

The Aryabhatiya contains exact formulas for the sum of the first n numbers, their squares or their cubes; we also find there the formula giving the area of a triangle or a circle as well as the approximate value $62\,832/20\,000$ for the π.

The book also includes a systematic treatment of the position of planets in space, gives an excellent order of magnitude of the circumference of the Earth and suggests that the apparent rotation of the heavens comes from the axial rotation of the Earth. The use of the decimal system will be one of the vectors of the development of Arab mathematics in arithmetic and algebra. His fame has survived the centuries and in his homage, the first Indian satellite, launched on *April* 19, 1975, as well as a lunar crater, bear his name.

Christian Goldbach was a Russian scientist.

He was not strictly speaking a mathematician,

but by his contacts with several of them,

including Bernouilli and de Moivre, and especially by his correspondences with Léonard Euler, led him to conceive the famous following conjecture, which carries his name: Any even integer greater than 3 can be written as the sum of two prime numbers.

CHRISTIAN GOLDBACH

(1690 – 1764)

In 1742, Christian Goldbach conjectured, in a letter to Leonhard Euler, that any whole number is the sum of at most three prime numbers. Harald Helfgott has just proven that this is true for odd numbers.

Leonhard Euler

atement of t ecture:

Any whole number strictly greater than 1 is the sum of at most three prime numbers.

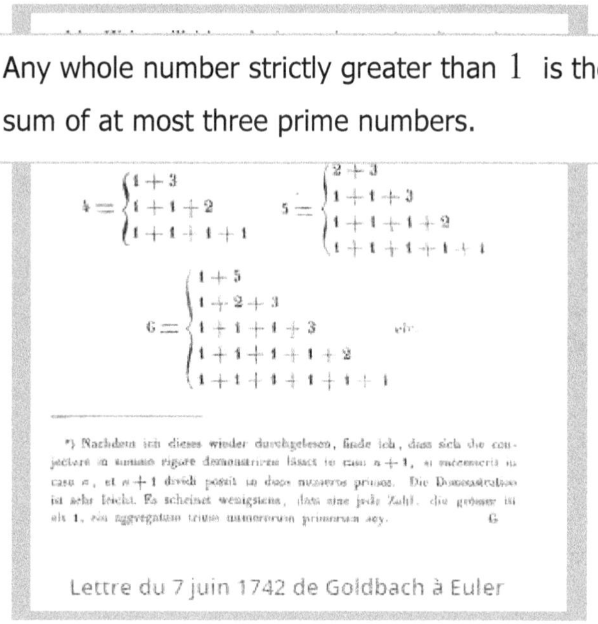

$$4 = \begin{cases} 1+3 \\ 1+1+2 \\ 1+1+1+1 \end{cases} \qquad 5 = \begin{cases} 2+3 \\ 1+1+3 \\ 1+1+1+2 \\ 1+1+1+1+1 \end{cases}$$

$$6 = \begin{cases} 1+5 \\ 1+2+3 \\ 1+1+1+3 \\ 1+1+1+1+2 \\ 1+1+1+1+1+1 \end{cases} \quad \text{etc.}$$

*) Nachdem ich dieses wieder durchgelesen, finde ich, dass sich die conjecture in summo rigore demonstriren lässet in casu $n+1$, si successerit in casu n, et $n+1$ dividi possit in duos numeros primos. Die Demonstration ist sehr leicht. Es scheinet wenigstens, dass eine jede Zahl, die grösser ist als 1, ein aggregatum trium numerorum primorum sey. G.

Lettre du 7 juin 1742 de Goldbach à Euler

Auguste De Morgan was a British mathematician and logician born in India. De Morgan lost his father in his tenth year. With his mother who wanted to make his son a religious, they lived in various cities in the south-west of England, which made him change schools often. His mathematical talents were discovered when he was fourteen. A family friend found him making an elaborate diagram of Euclid with a ruler and compasses and explained his logic to him. He studied law in London, where a new university, University College, is being created. This establishment intends to welcome in particular those who are not Anglican or Presbyterian. De Morgan then became a professor of mathematics. He is the founder with Boole of modern logic; in particular, he formulated the laws known as De Morgan's Laws.

AUGUSTE DE MORGAN

(1806–1871)

Auguste De Morgan is recognized for his rediscovery of the law of duality between sum and product, where the opposite of an aggregate (logical sum) is the compound (logical product) of the opposites of aggregators; the opposite of a compound is the aggregate of the opposites of the components.

De MORGAN's theorem is expressed by two relations:

$$\overline{A+B} = \overline{A}.\overline{B}$$
$$\overline{A.B} = \overline{A}+\overline{B}$$

German mathematician, philosopher, diplomat, jurist and philologist who has written in Latin, German and French. Some authors attribute the introduction of the words abscissa and ordered in 1692 to Leibniz.

Leibniz is interested in systems of equations and urges the use of determinants.

It is the source of several terms:

"Function" which he introduced in 1692

of the term "differential" (which Newton calls "fluxion").

It also creates several new ratings:

(•) $y = f(x)$: introduced in 1692 at the same time as the term "function"

(•) ∂t : for the differential

(•) $\int_{x_0}^{x} f(t).\partial t$: for the integral

GOTTFRIED LEIBNIZ

(1646 – 1716)

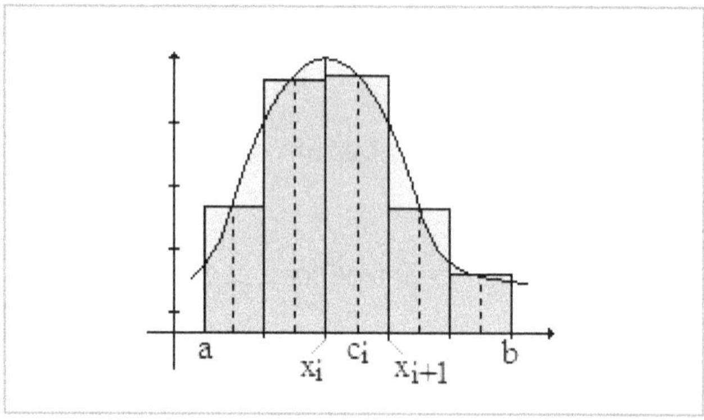

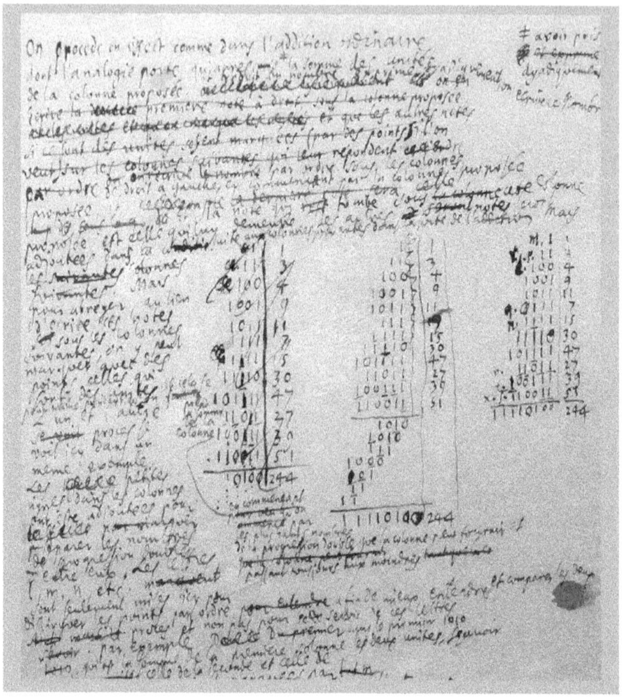

German mathematician, astronomer and physicist, whom some, like the King of Hanover, have given the nickname "prince of mathematicians".

Gauss's work is immense, to the point that legends, often unverified, have developed around his youth and his works. We often quote the anecdote of his solution to the problem of finding the sum of all integers from 1 to 100 . Despite having published very little during his life, we owe him a conjecture on the distribution of prime numbers, the complete classification of constructible polygons with ruler and compass, the fundamental theorem of algebra, an introduction to modular arithmetic, a description of non-Euclidean geometries, and several other results in mathematics, astronomy and physics.

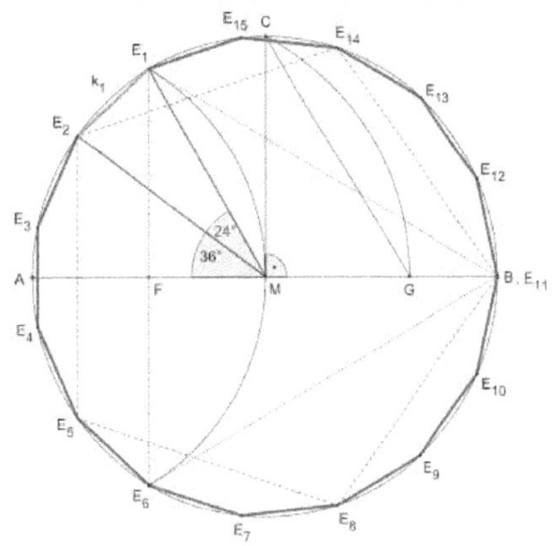

CARL FRIEDRICH
GAUSS

(1777 – 1855)

In the area of probability that the name of Gauss remains the most famous. He designed a continuous statistical law, called normal law or Laplace-Gauss law, whose distribution is represented by the famous bell curve.

The objective "normal" is explained by the fact that this law describes and models concrete and natural random statistical situations.

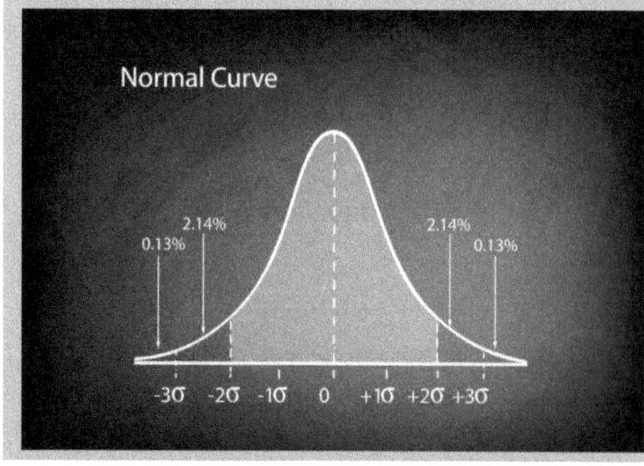

French mathematician who was the first to use factorial notation $(n!)$ As we know it today, in his work "Elements of Universal Arithmetic" *in* 1808. In fact, the more general concept of factorial was found at the same time by another French mathematician, Louis François Antoine Arbogast $(1759-1803)$.

$$1^{m|1} = 1.2.3 \ldots\ldots m$$

ou à cette autre forme plus simple

$$m!,$$

si l'on veut adopter la notation dont j'ai fait usage dans mes *Élémens d'arithmétique universelle*, n.° 289.

On a, en effet

$$a^{m|r} = a(a+r)(a+2r)\ldots\ldots[a+(m-1)r]$$
$$= r^m.\frac{a}{r}\left(\frac{a}{r}+1\right)\left(\frac{a}{r}+2\right)\ldots\left[\frac{a}{r}+(m-1)\right]$$
$$= r^m.\frac{1.2.3\ldots\left(\frac{a}{r}-1\right)\frac{a}{r}\ldots\left[\frac{a}{r}+(m-1)\right]}{1.2.3\ldots\left(\frac{a}{r}-1\right)}$$
$$= \frac{\left(\frac{a}{r}+m-1\right)!}{\left(\frac{a}{r}-1\right)!}r^m.$$

et

$$a^{m|-r} = a(a-r)(a-2r)(a-3r)\ldots\ldots[a-(m-1)r]$$
$$= r^m.\frac{a}{r}\left(\frac{a}{r}-1\right)\left(\frac{a}{r}-2\right)\ldots\left[\frac{a}{r}-(m-1)\right]$$

Nous terminerons par un rapprochement entre les notations de Vandermonde et celle de M. Kramp. Vandermonde fait

$$a.(a-1)(a-2)\ldots\ldots(a-m+1) = [\overset{m}{a}],$$

d'où il suit qu'en rapprochant les deux notations, on a

$$[\overset{m}{a}] = (a-m+1)^{m|1} = a^{m|-1}.$$

Si, après avoir changé a en a', on pose $a'-m+1=a$, d'où $a'=a+m-1$, on obtiendra cet autre rapprochement

$$[\overset{m}{a+m-1}] = a^{m|1} = (a+m-1)^{m|-1}.$$

Toutes les facultés pouvant être exprimées en fonction d'autres facultés dans lesquelles la base et la différence sont également l'unité, et ces dernières devant, en conséquence, se représenter fréquemment dans les calculs: M. Kramp, dans son *Arithmétique universelle*, a proposé de les écrire simplement comme il suit:

$$1.2.3.4\ldots\ldots m = 1^{m|1} = m!.$$

J. D. G.

CHRISTIAN KRAMP

(1760 – 1826)

" I use the very simple notation $n!$ to designate the product of numbers decreasing from n to unity, i.e.

$$n(n-1)(n-2)\ldots 3.2.1$$

The constant use in combinatorial analysis, in most of my proofs, that I make of this idea, has made this notation necessary "

Kramp's function, a scaled complex error function, is today better known as the Faddeeva function.

Christoff Rudolff was a German mathematician who was the first to use the symbol of the radical $(\sqrt{\ })$ in his work :

" Behend und hübsch Rechnung durch die kunstreichen regeln Algebra so gemeincklich die Coss genent werden " , printed *in* 1525 in Strasbourg. We therefore owe him the modern sign of the radical without the upper bar (vinculum). It was the mathematician René Descartes who added the horizontal bar to the symbol of Christoff in his work in Geometry, published *in* 1637.

Rudolff was aware that the development of mathematics depended on symbolism and notations. Therefore, he used notation as concise as possible.

CHRISTOFF RUDOLFF

(1499 – 1545)

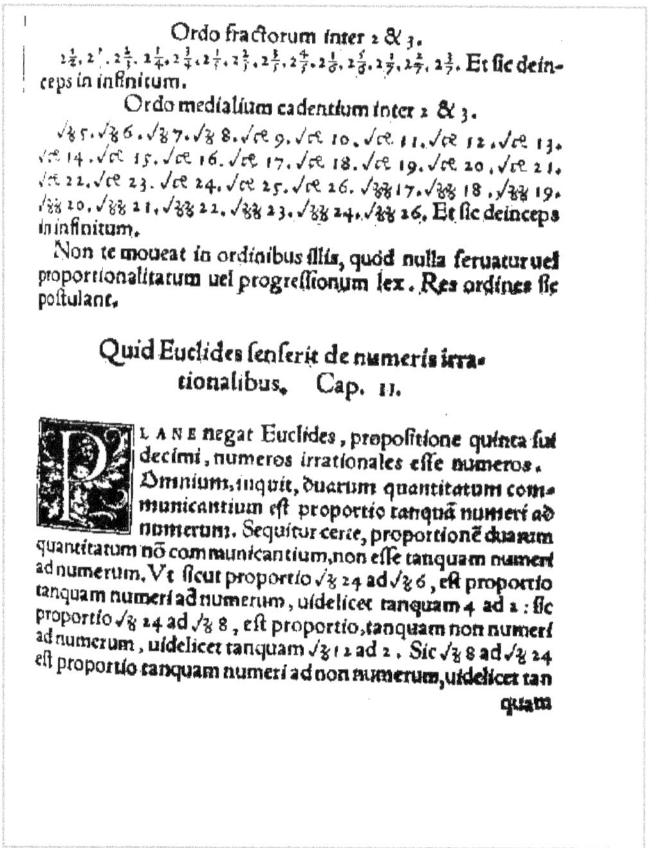

Ordo fractorum inter 2 & 3.

$2\frac{1}{2}, 2\frac{1}{3}, 2\frac{2}{3}, 2\frac{1}{4}, 2\frac{3}{4}, 2\frac{1}{5}, 2\frac{2}{5}, 2\frac{3}{5}, 2\frac{4}{5}, 2\frac{1}{6}, 2\frac{5}{6}, 2\frac{1}{7}, 2\frac{2}{7}, 2\frac{3}{7}$. Et sic deinceps in infinitum.

Ordo medialium cadentium inter 2 & 3.

√ʒ 5. √ʒ 6. √ʒ 7. √ʒ 8. √ce 9. √ce 10. √ce 11. √ce 12. √ce 13. √ce 14. √ce 15. √ce 16. √ce 17. √ce 18. √ce 19. √ce 20. √ce 21. √ce 22. √ce 23. √ce 24. √ce 25. √ce 26. √ʒʒ 17. √ʒʒ 18. √ʒʒ 19. √ʒʒ 20. √ʒʒ 21. √ʒʒ 22. √ʒʒ 23. √ʒʒ 24. √ʒʒ 26. Et sic deinceps in infinitum.

Non te moueat in ordinibus illis, quod nulla seruatur uel proportionalitatum uel progressionum lex. Res ordines sic postulant.

Quid Euclides senserit de numeris irrationalibus. Cap. II.

PLANE negat Euclides, propositione quinta sui decimi, numeros irrationales esse numeros. Omnium, inquit, duarum quantitatum communicantium est proportio tanquã numeri ad numerum. Sequitur certe, proportionẽ duarum quantitatum nõ communicantium, non esse tanquam numeri ad numerum. Vt sicut proportio √ʒ 24 ad √ʒ 6, est proportio tanquam numeri ad numerum, uidelicet tanquam 4 ad 2: sic proportio √ʒ 24 ad √ʒ 8, est proportio, tanquam non numeri ad numerum, uidelicet tanquam √ʒ 12 ad 2. Sic √ʒ 8 ad √ʒ 24 est proportio tanquam numeri ad non numerum, uidelicet tan quam

Scottish mathematician, theologian, physicist and astronomer. He popularized the use of the point for the Anglo-Saxon notation of decimal numbers and he invented logarithms.

In 1563, John Napier began his studies at the age of 13 at the University of St. Andrews. At first, he was passionate about theology.

It is believed that he left St. Andrews for the continent where he developed his knowledge in mathematics and classical literature, perhaps at the University of Paris or even in Italy and the Netherlands. After 1574 Napier and his wife resided in Scotland. He devoted himself to the management of his properties and used his inventive genius there. He took a scientific approach to agriculture.

JOHN NAPIER

$(1550-1617)$

In the preface to "Mirifici logarithmorum canonis descriptio", Napier said: "that he hopes that his logarithms will save calculators a lot of time and avoid errors in their

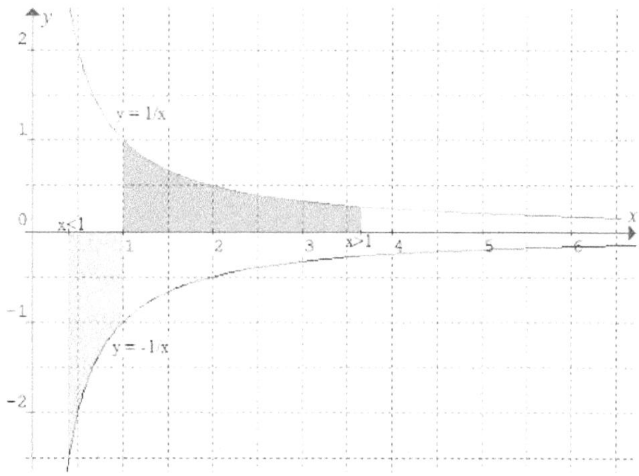

British theologian, mathematician and logician.

In his courses in logic and mathematics at the University of Cambridge, he took up the representation of the sets and syllogisms used by Euler and Carroll by replacing Euler's circles with less rigid curves and thus allowing the use of a model single simple to illustrate all possible attributes. He introduced the hatching to identify the empty areas.

His diagrams, renamed Venn diagrams, were widely popularized in the 1960s when mathematics programs called "modern mathematics" were revised.

JOHN VENN

(1834 – 1923)

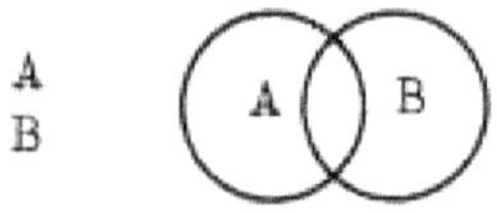

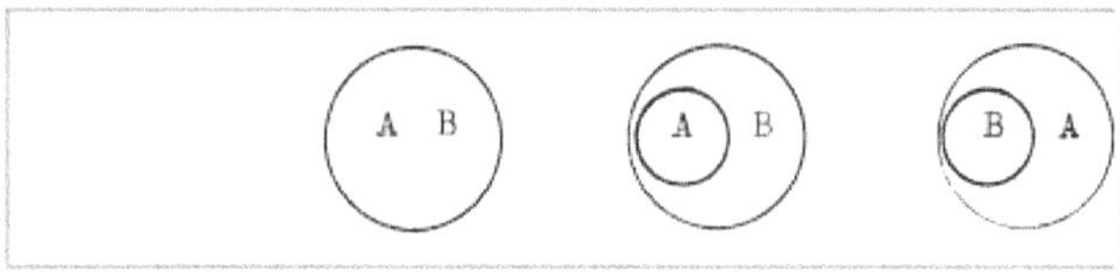

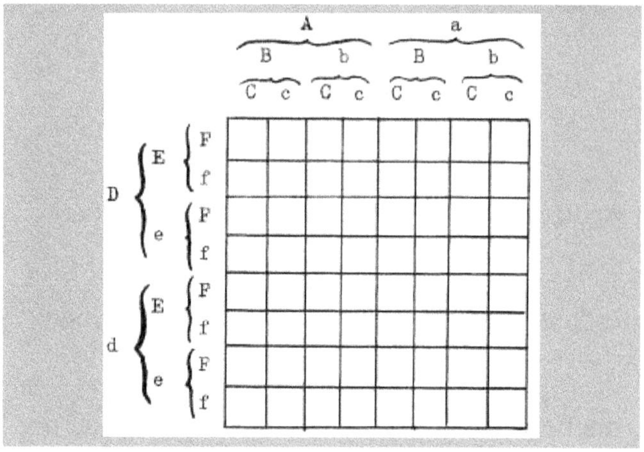

Charles Lutwidge Dodgson is best known today by his pseudonym Lewis Carroll, author of the tales The Adventures of Alice in Wonderland and On the Other Side of the Looking Glass. Dodgson was therefore a novelist, an essayist, a photographer and a math teacher. He was born *January* 27, 1832 in Daresbury, Cheshire, England, and died *January* 14, 1898 in Guildford. Following the work of Cayley (1821–1895), he worked on the concept of matrix, enriching it with many new and complementary results.

LEWIS CARROLL

(1832 – 1898)

Lewis Carroll's paradox:

By moving the four pieces that make up the square, we can arrange them so as to obtain the rectangle. Now the area of the square is 64 , while that of the rectangle is 65 . Where does the difference come from?

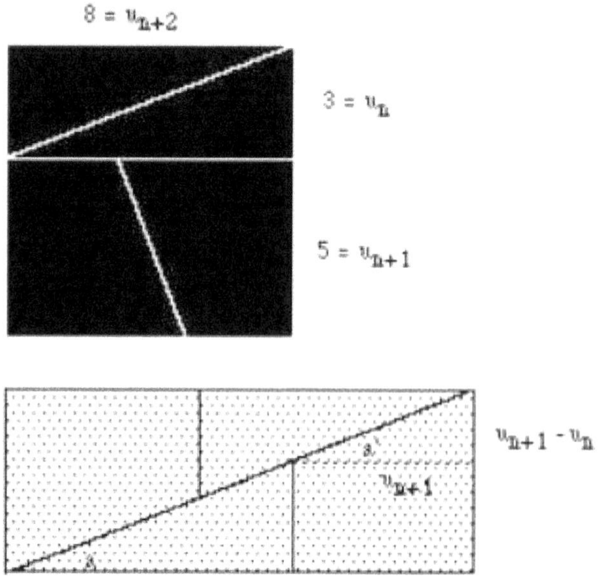

Paradox of Achilles and the tortoise.

German mathematician who proposed an axiomatic definition of the set of whole numbers as well as a rigorous construction of real numbers from rational numbers. The IN is due to Richard Dedekind who, $in\ 1888$, used it to designate the set of non-zero natural numbers. Today, this set is $IN*$. He also proposed the use of the symbol IR to denote the set of real numbers and the use of the symbol Z to denote the set of whole numbers.

Dirichlet takes a step in 1829 towards his general definition of an "arbitrary" function, which he will publish in 1837, giving the first example of a "totally discontinuous" function f defined by $f(x)=0$ if x is rational and $f(x)=1$ if x is irrational. On this occasion, he poses the question of the integration of bounded functions having an infinity of discontinuities in a finite interval, implicitly introducing the notion of rare set.

RICHARD DEDEKIND

(1831–1916)

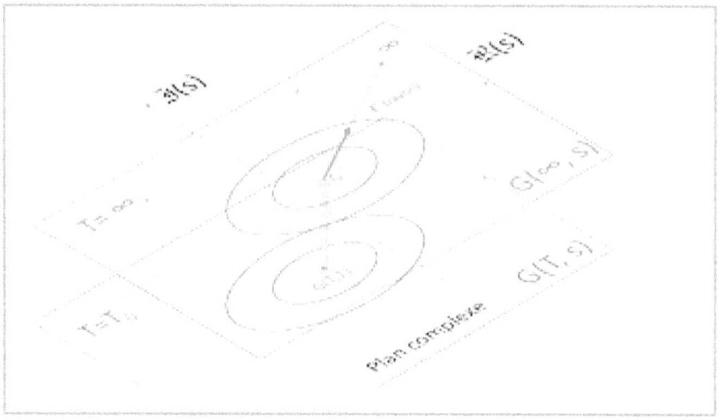

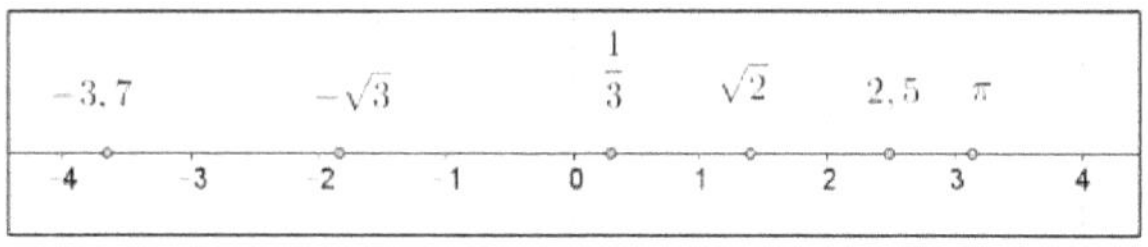

Pierre de Fermat, French mathematician, who contributed with Descartes to the creation of analytical geometry (he is the first to give a general method for the determination of tangents to a plane curve), to that of infinitesimal calculus (with Leibniz and Newton), and that of calculating probabilities (with Pascal). It is above all the founder of modern number theory, the branch of mathematics that studies whole numbers.

Among Fermat's problems, the postulate which consists in finding squares which are sums of two other squares (This is called looking for Pythagorean triplets, because they are the sides of a right triangle, for example: $5^2 = 3^2 + 4^2$)

Fermat's theorem was fully solved by the English mathematician Andrew Wiles in *September* 19, 1994.

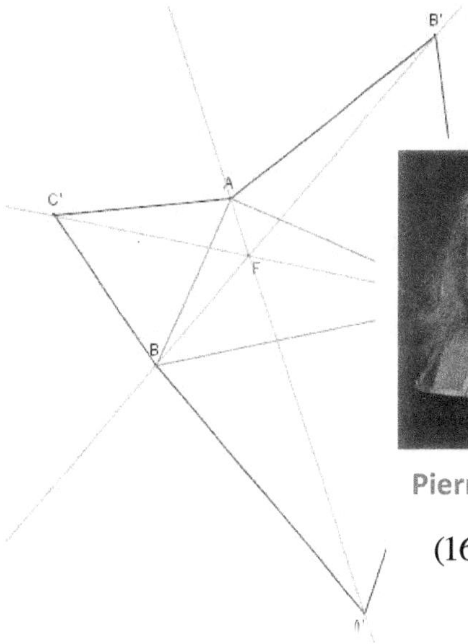

Pierre de Fermat

(1607 – 1665)

ermat wrote :

"On the other hand, a cube is never the sum of two cubes, a fourth power is never the sum of two fourth powers, and more generally no power greater than 2 is the sum of two analogous powers. found a wonderful demonstration of this proposition, but I cannot write it in this margin because it is too long "

Fermat's great theorem :

Let $n \geq 3$ be an integer. The solutions of $x^n + y^n = z^n$, where x, y and z are integers, all check $xyz = 0$

Fermat's little theorem :

Let p be a prime number, and x a prime integer with p . Then x^{p-1} has the remainder 1 in the division by p : $x^{p-1} \equiv 1 \ mod \ p$

Pell-Fermat equation :

We call the Pell-Fermat equation the diophantine equation $y^2 - tx^2 = 1$, where t is an integer which is not a square. The adjective "diophantine" means that we are looking for solutions with x and y which are integers.

Series to Riemann's hypothesis:

The Riemann series is the general term series:

$U_n = \frac{1}{n^{\alpha}}$; $\alpha \in IR$ It converges if $\alpha \succ 1$, and

Georg Friedrich Bernhard Riemann, born *September* 17,1826 in Breselenz, Kingdom of Hanover, died *July* 20, 1866 in Selasca, hamlet in the town of Verbania, Italy, is a German mathematician. Theoretically influential, he made many important contributions to topology, analysis, differential geometry and computation,

Riemann opened a new era in mathematics and his name is attached to a number of important concepts : the Riemann integral, the Cauchy-Riemann conditions, the Riemann surfaces, the Riemann function and the associated hypothesis, the theorem of Riemann's conformal application, Riemann Lebesgue's lemma.

Bernhard
Riemann
$(1826-1866)$

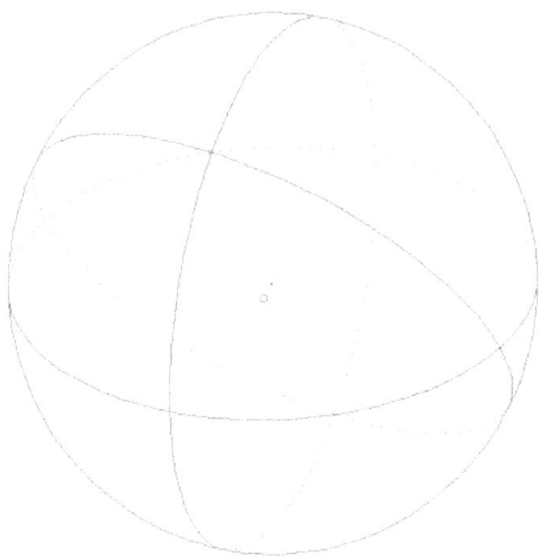

Des droites en géométrie riemannienne

$$\int_{\partial K^+} P\,dx + Q\,dy = \iint_K \left(\frac{\partial Q}{\partial x}(x,y) - \frac{\partial P}{\partial y}(x,y)\right) dxd$$

$$\begin{cases} \dfrac{\partial P}{\partial x} = \dfrac{\partial Q}{\partial y} \\ \dfrac{\partial P}{\partial y} = -\dfrac{\partial Q}{\partial x} \end{cases}$$

$$f(x,y) = P(x,y) + iQ(x,y)$$

Augustin-Louis Cauchy

(1789–1857)

Augustin-Louis Cauchy is the most prolific French mathematician (with almost 800 articles published).

He held various positions in Paris, at the Faculty of Sciences, and at the Polytechnic school. In 1816 he won the grand prize of the Academy of Sciences.

Cauchy was a pioneer in the study of the analysis and theory of substitution groups (now called permutations). Cauchy proved in 1811 that the angles of a convex polyhedron are determined by its faces. In 1814 he published a dissertation on definite integrals which became the basis of the theory of complex functions.

Many mathematical terms bear his name:

(•) Cauchy's integral theorem (in theory of complex functions)

(•) The Cauchy-Kovalewskaya theorem (for solving partial differential equations).

(•) The Cauchy-Riemann equations.

Cauchy Integral Formula :

$$f(a)=\frac{1}{2\pi i}\int_C \frac{f(z)}{z-a}\,dz$$

Cauchy's Law :

A random variable X follows Cauchy's law if it is absolutely continuous and admits for density:

$$f_X(x)=\frac{1}{\pi}\times\frac{1}{1+x^2} \quad , \ \forall x\in IR$$

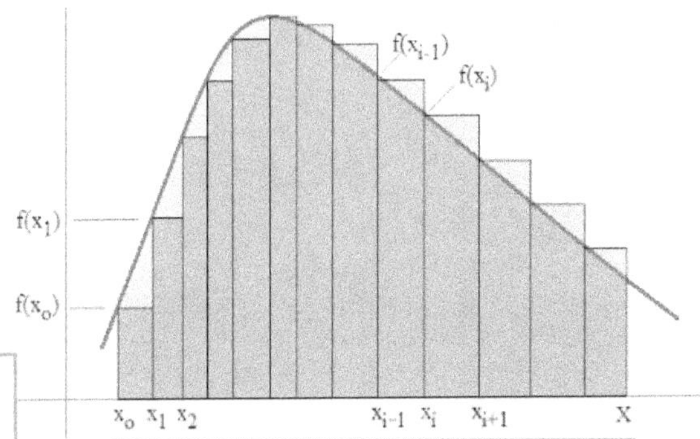

$$S(f,\sigma)=\sum_{i=1}^{n} f(x_{i-1})(x_i-x_{i-1})$$

Henri Poincaré is a French mathematician, physicist, philosopher and engineer, born *April* 29, 1854 in Nancy and died *July* 17, 1912 in Paris.

Poincaré has carried out major work in optics and infinitesimal calculus.

He is also a major precursor to the theory of special relativity, the theory of dynamical systems, and the theory of chaos.

Poincaré has shown that there is no analytical solution to the problem of the three bodies.

He states the properties of the Poincaré - Lorenz group, which will lead to Einstein's fundamental article on special relativity.

His conjecture on topology will occupy mathematicians for a century. It was resolved *in* 2002

Henri Poincaré

(1854 – 1912)

" The real method of predicting the future of mathematics is the study of its history and current state "

" The more science expands the circle of knowledge and the more the circle around the shadow grows "

" Doubting everything or believing everything are two equally convenient solutions which both make us think"

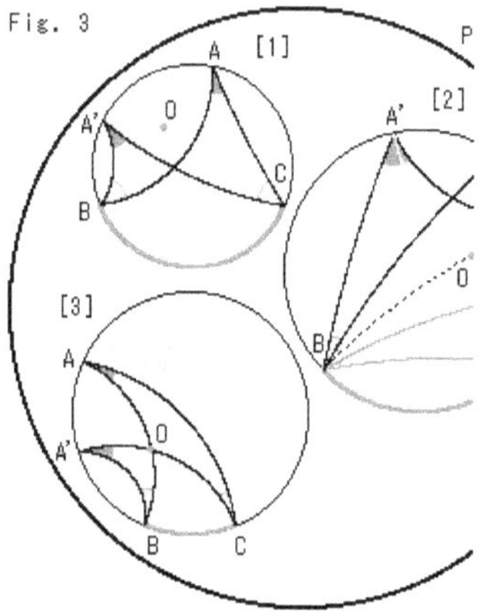

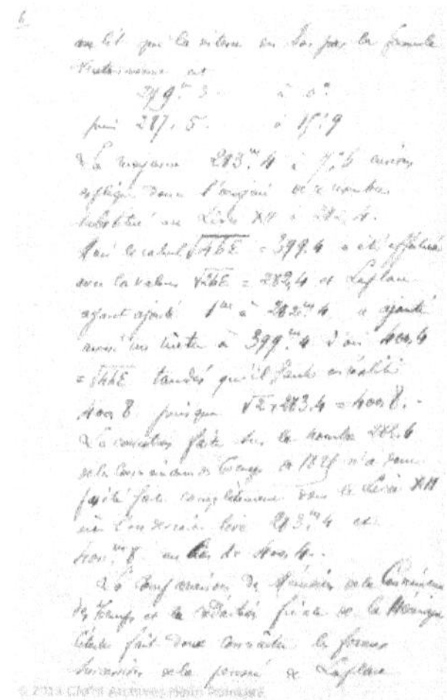

Isaac Newton was born in Woolsthorpe on *December* 25, 1642.

Newton is interested in astronomy, and mathematics because he lacks many geometric concepts, to understand the work of Halley.

Newton understands that white light is not an entity, but is the sum of colored lights.

In mechanics, he established the three universal laws of movement which in fact constitute principles at the base of Newton's great theory concerning the movement of bodies, a theory that we now call "Newtonian mechanics" or even "mechanics classic ".

It is also known for the generalization of the binomial theorem and the invention known as the Newton method allowing to find approximations of a zero of a function of a real variable with real values.

Newton's binomial - Binomial coefficients :

If a and b are two complex numbers, and n is a strictly positive integer:

$$(a+b)^n = \sum_{k=0}^{n} \binom{n}{k} a^k b^{n-k}$$

Isaac Newton

$(1642 - 1727)$

Newton's trident is the equation curve: $xy = ax^3 + bx^2 + cx + d$ According to the values of the coefficients, Newton's trident cuts the abscissa axis in one or three points, and if d is not zero, the ordinate axis is asymptote to the curve.

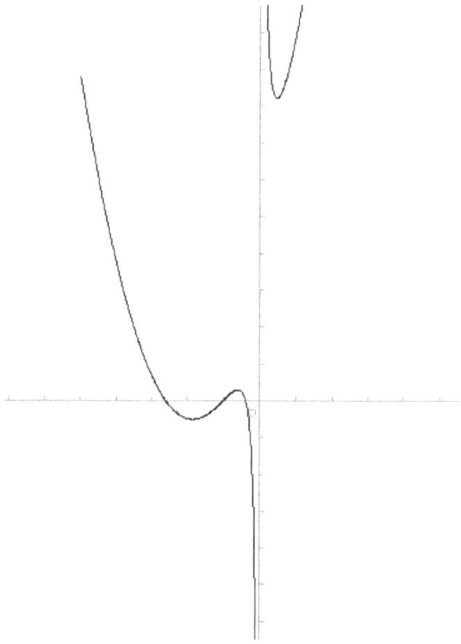

Newton's Trident

PART 02

Developing Conjectures

PART 02

A conjecture is a mathematical statement that has not yet been rigorously proved. Conjectures arise when one notices a pattern that holds true for many cases. However, just because a pattern holds true for many cases does not mean that the pattern will hold true

Developing Conjectures

Pascal's triangle

The 0th through 4th rows of Pascal's triangle are

Developing Conjectures

SEGMENTS CONNECTING

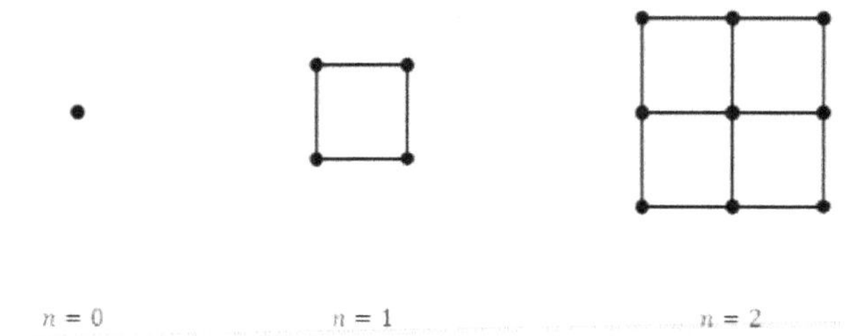

Let X_n be the number of segments that connect an $n \times n$ square lattice. Conjecture an expression for X_n

Observing the cases given by counting how many segments are in each figure, we have :

$$X_0 = 0$$
$$X_1 = 4$$
$$X_2 = 12$$

From these three cases $n = 3$

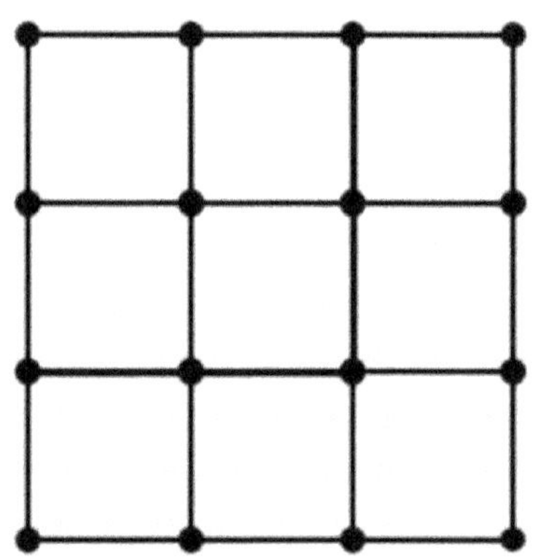

Counting the segments here gives $X_3 = 24$

One might notice that each difference between consecutive terms in the sequence is a multiple of 4 :

$$X_1 - X_0 = 4$$
$$X_2 - X_1 = 8$$
$$X_3 - X_2 = 12$$
$$X_4 - X_3 = 16$$

Recurrence relation

$$(\forall n \in IN): X_n = X_{n-1} + 4n$$

It would be more desirable to develop an expression for X_n purely in terms of n

A better way to tackle these kinds of problems is to think more creatively about how the problem is structured.

Developing Conjectures

SEGMENTS CONNECTING

For $n=3$, except now the horizontal and vertical segments are color-coded :

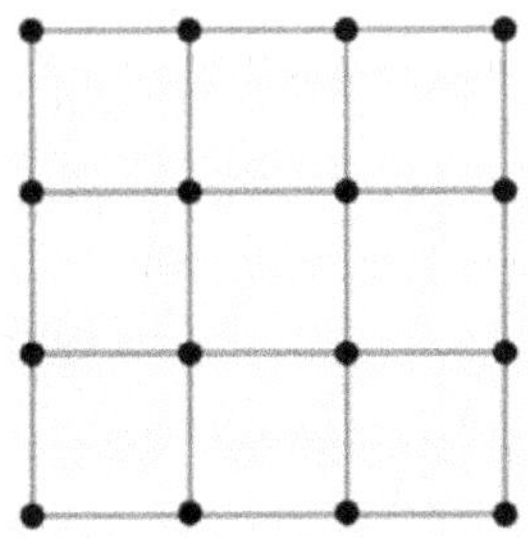

Notice that there are 4 horizontal lengths, and each of them consists of 3 segments.The same is true for vertical lengths.

So, the total number of segments is:

$$X_3 = 2(3)(4) = 24$$

Consider the other cases:

$$X_0 = 2(0)(1)=0$$
$$X_1 = 2(1)(2)=4$$
$$X_2 = 2(2)(3)=12$$

CONJECTURE

The number of segments connecting an $n \times n$ lattice is defined by the sequence

$$(\forall n \in IN): X_n = 2n(n+1)$$

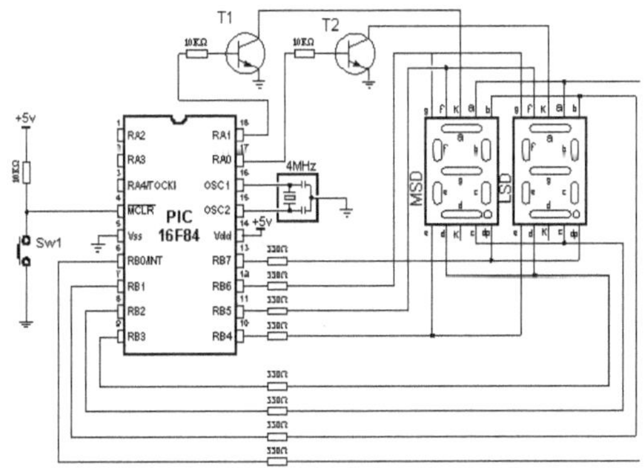

Quelle: Deutsche Fotothek

Developing Conjectures

PARTITION FUNCTION

A partition of a positive n is an expression of n as the sum of one or more positive integers (or parts). The order of the integers in the sum "does not matter": that is, two expressions that contain the same integers in a different order are considered to be the same partition.

The number of different partitions of n is $p(n)$. This function is called the partition function.

$$p(2)=2 \quad ; \quad p(3)=3$$
$$p(4)=5 \quad ; \quad p(5)=7$$
$$p(6)=11 \quad ; \ldots..$$

A partition of a positive integer n is a non-increasing sequence of positive integers whose sum is n.The partition function $p(n)$ is defined to be the number of partitions of n. By convention $p(0)=1$ and $p(n)=0$ for $n<0$

Euler showed that the partition function satisfies the following generating function relationship:

$$\sum_{n=0}^{\infty} p(n)q^n = \prod_{n=1}^{\infty} \frac{1}{1-q^n}$$

An $r-colored$ partition of a positive integer n is a partition of n, where one of r *colors* is assigned to each integer in the sequence. r^{th} power of the partition function, $p_r(n)$, counts the number of $r-colored$ partitions of n. It satisfies a generating function relationship similar to that of $p(n)$:

$$\sum_{n=0}^{\infty} p_r(n)q^n = \prod_{n=1}^{\infty} \frac{1}{(1-q^n)^r}$$

CONJECTURE

The number of partitions of an integer n is p_{n-1}, Where p_k is the kth prime number

Observing the very next value of $p(n)$ puts this conjecture to rest: $p(7)=15$. As soon as a single case is shown to disobey the pattern, the conjecture is disproved. This is called a counterexample. Once a

Open Conjectures

There are many open conjectures in mathematics. An open conjecture is one that has been proposed, but no formal proof has yet been developed

GOLDBACH'S CONJECTURE

The Goldbach Conjecture is a yet unproven conjecture stating that every even integer greater than two is the sum of two prime numbers. The conjecture has been tested up to

$$4\times10^{18}$$

Goldbach's conjecture is one of the oldest unsolved problems in number theory and in all of mathematics.

For example :

$$4=2+2$$
$$6=3+3$$
$$8=3+5$$
$$10=3+7=5+5$$
$$12=5+7$$
$$14=3+11=7+7$$

The former conjecture is today known as the "ternary" Goldbach conjecture, the latter as the "strong" or "binary" Goldbach conjecture. The

CONJECTURE

Every integer greater than 2 can be written as the sum of three primes

Version of the conjecture by **Euler**

CONJECTURE

Every even number greater than 2 can be written as the sum of two primes

Open Conjectures

GOLDBACH'S CONJECTURE

The majority of mathematicians believe the conjecture (in both the weak and strong forms) to be true, at least for sufficiently large integers, mostly based on statistical considerations focusing on the probabilistic distribution of prime numbers: the bigger the number, the more ways there are available for that number to be represented as the sum of two or three other numbers, and the more "likely" it becomes that at least one of these representations consists entirely of primes.

A very crude version of the heuristic probabilistic argument (for the strong form of the Goldbach conjecture) is as follows :

An integer m selected at random has roughly $\frac{1}{\ln(m)}$ chance of being prime.

This heuristic is non-rigorous for a number of reasons: for instance, it assumes that the events that m and $(n-m)$ are prime are statistically independent of each other. Nevertheless, if one pursues this heuristic, one might expect the total number of ways to write a large even integer n as the sum of two odd primes to be roughly :

$$\sum_{m=3}^{n/2}\frac{1}{\ln(m)}\frac{1}{\ln(n-m)}\approx\frac{n}{2\ln^2(n)}$$

Since this quantity goes to infinity as n increases, we expect that every large even integer has not just one representation as the sum of two primes, but in fact has very many such

Open Conjectures

TWIN PRIME CONJECTURE

Proposed 1849 by **Alphonse de Polignac**

CONJECTURE

The Twin Prime Conjecture is a conjecture that states that there are infinitely many pairs of twin primes, i.e. pairs of primes that differ by 2

The first couple of twin prime pairs are $(3,5);(5,7)$ and $(11,13)$. Larger and larger pairs of twin primes continue to be discovered, as of *September* 2016, the largest known twin prime pair is

$$2996863034895 \times 2^{1290000} \pm 1$$

One possible strategy to prove the infinitude of twin primes is an idea adopted from the proof of Dirichlet's Theorem. If one can show that the sum $\Phi = \frac{1}{3}+\frac{1}{5}+\frac{1}{7}+\frac{1}{11}+\frac{1}{13}+....$, of the reciprocals of twin primes diverges, this would imply that there are infinitely

A weaker version of twin prime conjecture was proved by Yitang Zhang in 2013. This version stated that there are infinitely many pairs of primes that differ by a finite number.

Open Conjectures

Riemann Hypothesis

Proposed 1859 by **Bernhard Riemann**

The Riemann hypothesis is one of the most important open problems in mathematics. If it were to be proved, it would lead to several important developments in number theory and algebra. The most notable of these potential developments would be a better understanding of the distribution of primes.

The Riemann Hypothesis is a famous conjecture in analytic number theory that states that all nontrivial zeros of the Riemann zeta function have real part $\frac{1}{2}$.

From the functional equation for the zeta function, it is easy to see that $\zeta(s)=0$ when $s=-2,-4,-6,.....$.

These are called the trivial zeros. This hypothesis is one

The Riemann Hypothesis is an important problem in the study of prime numbers. Let $\pi(x)$ denote the number of primes less than or equal to x, and let $L_i(x)=\int_2^x \frac{1}{\ln(y)}dy$ Then an equivalent statement of the Riemann hypothesis is that $\pi(x)=L_i(x)+o\left(\sqrt{x}\ln(x)\right)$.

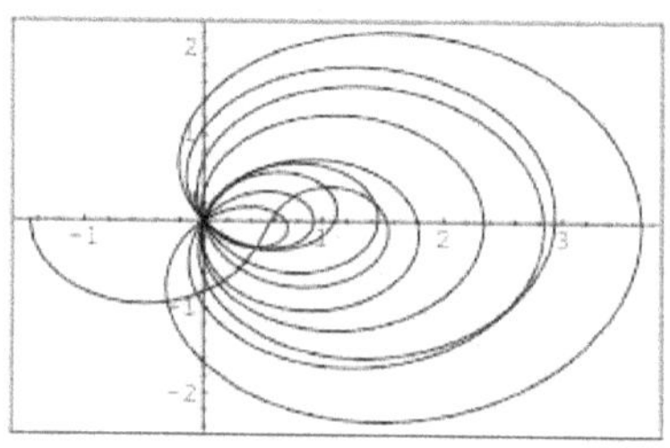

Figure 1. $\zeta(\frac{1}{2}+it)$ for $0<t<50$.

If I were to awaken after having slept for a thousand years, my first question would be: Has the Riemann hypothesis been proven?

David Hilbert

The Riemann zeta function for $s \in C$ with $\mathrm{Re}(s) \succ 1$ is defined as $\zeta(s) = \sum_{n=1}^{\infty} \frac{1}{n^s}$ It is then defined by analytical continuation to a meromorphic function on the whole C by a function equation

Open Conjectures

Riemann Hypothesis

Proposed 1859 by **Bernhard Riemann**

The Riemann zeta function for $s \in C$ with $\mathrm{Re}(s) \succ 1$ can be written as :

$$\zeta(s) = \prod_{p\,prime} \frac{1}{1-p^{-s}}$$

Euler Product Representation

Euler's proof

This proof will deal only with real number $s \succ 1$, By absolute convergence :

Recall the formula for the sum of a geometric series:

$$1 + q + q^2 + q^3 + ... = \frac{1}{1-q} = (1-q)^{-1} \ ; \quad (|q| \prec 1)$$

Letting $q = p^{-s}$, where p is a prime,

We obtain :

Open Conjectures

Riemann Hypothesis

Proposed 1859 by **Bernhard Riemann**

Where $S(p,q,...)$ denotes the set of all such integers – Including $(1 = p^0 q^0 ...)$ - whose prime factorizations involve only prime in $\{p,q,...\}$, which is a set of primes.

When we multiply by $(1-5^{-s})^{-1}$, We get :

$$(1-2^{-s})^{-1}(1-3^{-s})^{-1}(1-5^{-s})^{-1} = \sum_{n \in S(2,3,5)} \frac{1}{n^s}$$

In general, if p_m is the mth prime,

We have :

> The Riemann hypothesis is a precise statement, and in one sense what it means is clear, but what it's connected with, what it implies, where it comes from, can be very unobvious

M. Huxley

Open Conjectures

Riemann Hypothesis

Proposed 1859 by **Bernhard Riemann**

The Riemann zeta function for $s \in C$ with $\mathrm{Re}(s) \succ 1$ can be written as :

$$\zeta(s) = \prod_{p\ prime} \frac{1}{1-p^{-s}}$$

Euler Product Representation

PROOF

Since $\frac{1}{n^s}$ is multiplicative, we take the product of the sum over all prime powers :

$$\sum_{n=1}^{\infty}\frac{1}{n^s}=\prod_{p=prime}\sum_{k=0}^{\infty}\frac{1}{\left(p^k\right)^s}$$

The LHS's sum is just the geometric progression sum. We have it equal to

$$\sum_{n=1}^{\infty}\frac{1}{n^s}=\prod_{p=prime}\frac{1}{1-p^{-s}}$$

The mathematician's patterns, like the painter's or the poet's, must be beautiful; the ideas, like the colours or the words must fit together in a harmonious way. Beauty is the first test: there is no permanent place in this world for ugly mathematics

G. H. Hardy, 1940

Open Conjectures

Riemann Hypothesis

Proposed 1859 by **Bernhard Riemann**

Integral Representation

The zeta function can be represented as : $\Gamma(s)\zeta(s)=\int_0^{\infty}\frac{x^{s-1}}{e^x-1}dx$, When $\Gamma(s)=\int_0^{\infty}x^{s-1}e^{-x}\,dx$

PROOF

Consider the gamma function : $\Gamma(s)=\int_0^\infty x^{s-1}e^{-x}\,dx$

We substitute $x=nt$ and thus $dx=ndt$: $\Gamma(s)=\int_0^\infty n^{s-1}t^{s-1}e^{-nt}n\,dt=\int_0^\infty n^s t^{s-1}e^{-nt}\,dt$

So, $\Gamma(s)\frac{1}{n^s}=\int_0^\infty t^{s-1}e^{-nt}\,dt$

We take the sum over all the positive integers as n :

$$\Gamma(s)\sum_{n=1}^{\infty}\frac{1}{n^s}=\sum_{n=1}^{\infty}\int_0^\infty t^{s-1}e^{-nt}\,dt=\int_0^\infty t^{s-1}\sum_{n=1}^{\infty}e^{-nt}\,dt$$
$$=\int_0^\infty t^{s-1}\frac{1}{e^t-1}\,dt$$
$$=\int_0^\infty \frac{t^{s-1}}{e^t-1}\,dt$$

So, $\Gamma(s)\sum_{n=1}^{\infty}\frac{1}{n^s}=\int_0^\infty\frac{t^{s-1}}{e^t-1}\,dt \quad\Rightarrow\quad \Gamma(s)\zeta(s)=\int_0^\infty\frac{t^{s-1}}{e^t-1}\,dt$

Open Conjectures

Mertens Conjecture

The Mertens conjecture states that $|M(x)|\prec\sqrt{n}$ for all $x>1$, where $M(x)=\sum_{n\leq x}\mu(n)$

and $\mu(n)$ is the Möbius function. This conjecture has attracted a substantial amount of interest in its almost 100 years of existence because its truth was known to

The Möbius μ function is defined from IN^* *in* $\{-1,\ 0,\ 1\}$. The image $\mu(n)$ of an integer $n>0$ is worth:

$(\bullet)$ 0 if n is divisible by a perfect square other than 1.

inequality of form $|M(n)| < c\sqrt{n}$, c being a positive real, implies the Riemann hypothesis :

$$\forall \varepsilon \succ 0, \quad M(x) = o\left(x^{1/2+\varepsilon}\right)$$

$$\Downarrow$$

$$\frac{1}{\zeta(s)} = s\int_1^\infty \frac{M(t)}{t^{z+1}}dt$$

where ζ is Riemann's zeta function

Odlyzko and te Riele, proved that Mertens' conjecture is false.

They showed that: $\limsup_{n\to\infty} M(n)n^{-1/2} \succ 1.06 \quad and \quad \liminf_{n\to\infty} M(n)n^{-1/2} \prec -1.009$

Open Conjectures

abc Conjecture

Proposed 1985 by **Joseph Oesterlé and David Masser**

The abc conjecture is a conjecture due to Oesterlé and Masser in 1985 . It states that, for any infinitesimal $\varepsilon > 0$, there exists a C_ε such

EXAMPLE

Let $a = 49$; $b = 75$; and $c = a + b$, let d be the product of distinct prime factors of abc. Show

Joseph Oesterlé

David Masser

Open Conjectures

Collatz conjecture

Proposed 1930 by **Lothar collatz**

Lothar Collatz likely posed the eponymous conjecture in the $1930s$. The problem sounds

Over the years, many problem solvers have been drawn to the beguiling simplicity of the Collatz conjecture, or the "$3x+1$ *problem*" as it's also known.

Consider the following operation on an arbitrary positive integer $U_0 = N$:

If the number is even, divide it by two.

If the number is odd, triple it and add one.

$$(\forall n \in IN):$$
$$\begin{cases} U_{n+1} = \frac{U_n}{2} & ; if\ U_n\ is\ even\ number \\ U_{n+1} = 3U_n + 1 & ; if\ U_n\ is\ odd\ number \end{cases}$$

In modular arithmetic notation, define the function Φ as follows:

$$\Phi(n) = \begin{cases} \frac{n}{2} & ;\ if\ n \equiv 0[2] \\ 3n+1 & ;\ if\ n \equiv 1[2] \end{cases}$$

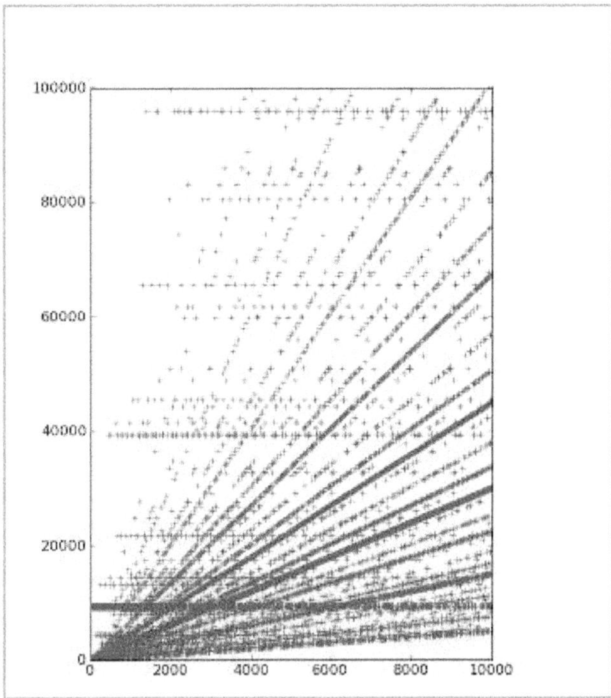
100000
80000
60000
40000
20000
0
0
2000
4000
6000
8000
10000

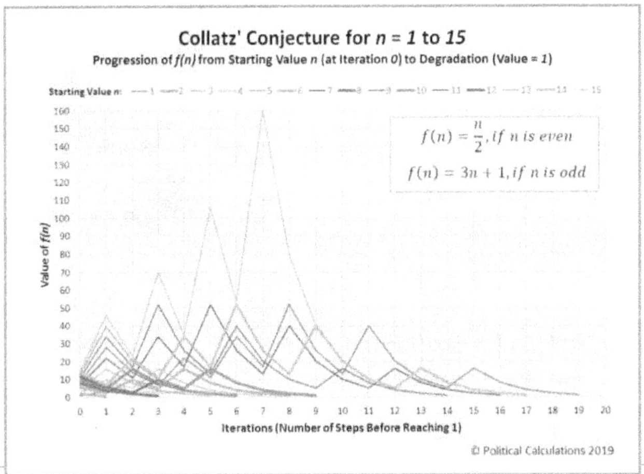
Collatz' Conjecture for n = 1 to 15
Progression of f(n) from Starting Value n (at Iteration 0) to Degradation (Value = 1)
Starting Value n:
f(n) = n/2, if n is even
f(n) = 3n + 1, if n is odd
Value of f(n)
Iterations (Number of Steps Before Reaching 1)
© Political Calculations 2019

Open Conjectures

Hadamard Conjecture

Proposed 1893 by **Jacques Hadamard**

This problem was formulated by the famous French mathematician Jacques Hadamard in 1893 , one year after his thesis and three years before his famous proof of the Prime Number Theorem.

Jacques Salomon HADAMARD

CONJECTURE

For any natural integer n multiple of 4, there is a Hadamard matrix of order n

The Hadamard Matrix H is a square $n \times n$ matrix, with entries ± 1, and the first row and the first column consist of $+1$ only, such that $HH^T = nI_n$. It is an orthogonal matrix, where the inner product of any two rows, which is denoted by g is 0.

The Hadamard Matrix H

$$\begin{pmatrix} 1 & 1 & 1 & 1 \\ 1 & -1 & 1 & -1 \\ 1 & 1 & -1 & -1 \\ 1 & -1 & -1 & 1 \end{pmatrix}$$

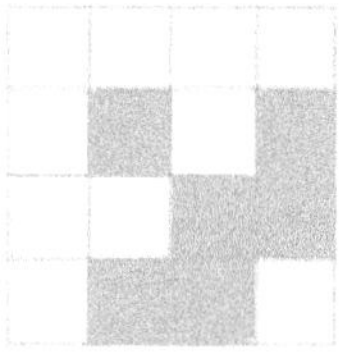

Example

These matrix were invented by **Sylvester** in 1867 , who built a power of 2 of any order (Anallagmatic chessboards)

James Joseph SYLVESTER (1814-1897)

Let M be a square matrix of order n, with real coefficients between -1 and 1. Then the determinant of M is between $\left(-n^{n/2}\right)$ and $\left(n^{n/2}\right)$. Furthermore, if $det(M)$ reaches one of these two bounds, then M is a Hadamard matrix, and vice versa.

Theorem (Hadamard, 1893)

Open Conjectures

Hadamard Conjecture

Proposed 1893 by **Jacques Hadamard**

The first mathematician to have effectively captured Hadamard's subtle message is **Umberto Scarpis** :

Umberto SCARPIS
(1861- 1921)

Theorem (1898)

Let n be a multiple of 4 such that $(n-1)$ is a prime number. If there is a Hadamard matrix of order n, then there is a larger one of order $(n-1)n$.

Ray Edwin GILMAN

(1887-1975)

Theorem (1930)

Let n be a multiple of 4 such that $(n-1)$ is a prime number. Then there is a Hadamard matrix of order n.

Theorem (1933)

Let n be a multiple of 4 such that $(n-1)$ or $(n/2-1)$ be a prime number power. Then there is a Hadamard matrix of orde n.

Raymond Edward Alan Christopher PALEY

(1907-1933)

Open Conjectures

Melham's Conjecture on Odd Power Sums of Fibonacci Numbers

Fibonacci numbers are represented by $\Im_n$ and follow the sequence: $1,1,2,3,5,8,13$ ..., where each term is the sum of the previous two, with the exception of the first two terms.

Thus, the Fibonacci sequence $\Im_0, \Im_1, \Im_2, \ldots$ can be entirely defined by the following formulas:

$$\begin{cases} \Im_0 = 0 \\ \Im_1 = 1 \\ \Im_n = \Im_{n-2} + \Im_{n-1} \quad ; n \geq 2 \end{cases}$$

This sequence has a multitude of properties which makes it the most studied numerical sequence, but the main property, in our opinion, is its relationship with the golden ratio, Φ. Let's check it by finding the proportion resulting from the division of one of these numbers by the previous one:

$$\frac{\Im_2}{\Im_1} = \frac{1}{1} = 1 \quad ; \quad \frac{\Im_3}{\Im_2} = \frac{2}{1} = 2$$

$$\frac{\Im_4}{\Im_3} = \frac{3}{2} = 1.5 \quad ; \quad \frac{\Im_5}{\Im_4} = \frac{5}{3} \simeq 1.6666$$

$$\frac{\Im_6}{\Im_5} = \frac{8}{5} = 1.6 \quad ; \quad \frac{\Im_7}{\Im_6} = \frac{13}{8} \simeq 1.625$$

$$\frac{\Im_8}{\Im_7} = \frac{21}{13} \simeq 1.61538$$

The more we advance, the closer we get to the golden ratio $\Phi = \dfrac{\sqrt{5}+1}{2} \simeq 1{,}6180339887\ldots$.

In fact, the limit of the sequence described above is the number of gold: $\lim\limits_{n \to +\infty} \dfrac{\Im_n}{\Im_{n-1}} = \Phi$

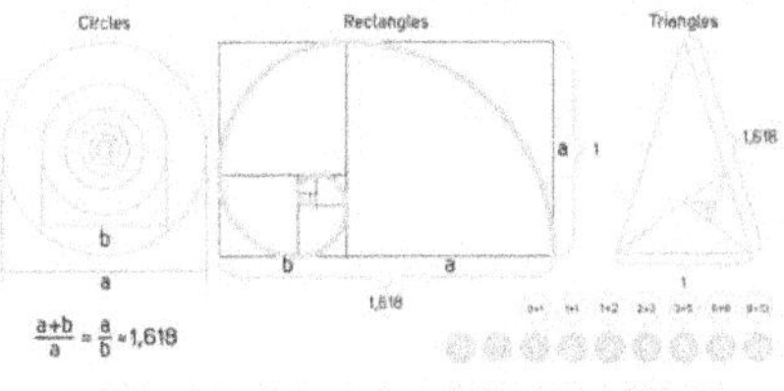

Open Conjectures

Melham's Conjecture on Odd Power Sums of Fibonacci Numbers

Let $\Im_n$ denote the nth Fibonacci number, and let L_n denote the nth Lucas number. It is well known that the Fibonacci numbers and the Lucas numbers satisfy the same recurrence relation

$$\Im_n = \Im_{n-1} + \Im_{n-2}; \qquad L_n = L_{n-1} + L_{n-2}; \qquad for\ n \geq 2$$

With different initial values $\Im_0 = 0,\ \Im_1 = 1,\ L_0 = 2\ \ and\ \ L_1 = 1$.

CONJECTURE

For any positive integers n, m, the sum

$\prod_{k'=0}^{m} L_{2k'+1} \sum_{k=1}^{n} \Im_{2k}^{2m+1}$ can be expressed as $(\Im_{2n+1} - 1)^2 P_{2m-1}(\Im_{2n+1})$, where $P_{2m-1}(X)$ is a polynomial of degree $2m-1$ with integer coefficients

Famous MATHEMATICIANS

Alan Turing was a mathematician and logician who also contributed in the fields of cryptanalysis and philosophy, while in later life contributing to computer science and artificial intelligence. Turing worked at the British Government's Code and Cypher School before World War II broke out, carrying out top secret work deciphering the military codes used by Germany and its allies.

Fibonnaci, or Leonardo Bonacci, was an Italian Mathematician who was considered to be "the most talented Western mathematician of the middle ages." He is best known for his work on the Fibonacci sequence. Before Fibonacci, Europe used Roman Numerals which made modern mathematics almost impossible. He advocated the use of the digits 0-9, and of place value.

Pythagoras was an Ancient Greek mathematician and philosopher. He is best known for the Pythagorean theory named after him. Often referred to as Pythagoras' Theorem, it states that in a right-angled triangle, the square on the hypotenuse is equal to the sum of the squares on the other two sides.

Charles Babbage was an English mathematician, philosopher, inventor and mechanical engineer. He is among the most famous icons in the pre-history of computing. Babbage's 'Difference Engine 1', was the first successful example of an automatic calculator, which remains one of the finest examples of engineering of all time.

Notation

The notation in this book is standard, Specific symbols and functions are defined as needed throughout, and the standard meaning of basic symbols ans function is assumed. The following is a list of symbols that appear frequently in the text, and the meanings we intend them to convey:

$\mathbb{N}$: *Natural numbers* $\{0,1,2,...\}$

$\mathbb{Q}$: *Rational numbers*

$\mathbb{Z}$: *The set of integers*

$\mathbb{Z}/p\mathbb{Z}$: *Integers mod p*

$\mathbb{R}$: *Real numbers*

$\mathbb{C}$: *The set of complex numbers*

$\sim$: *Similarity*

$\simeq$: *Approximately equal*

$\Rightarrow$: *Implication (if ..., then...)*

$\Leftrightarrow$: *Equivalence (if and only if)*

$\forall$:*Universal quantification (for all)*

$\exists$: *Existential quantification (there exists)*

$\exists!$:*Uniqueness quantification*

$|x|$: *The absolute value of* x

$\binom{k}{n}$: *Combination; binomial coefficient*

∂: *Partial derivative*

$\mathfrak{I}_n$:*Fibonacci numbers*

$\in$:*Set membership*

$\notin$:*Is not an element of*

$\equiv$: *Congruence relation*

$\sum$:*Summation*

$\prod$:*Product*

$\pi(x)$: *The number of prime numbers* $p \leq x$

ζ :*Riemann's zeta function*

Φ : *Golden ratio*

Γ :*Gamma function* $\Gamma(t)=\int_0^\infty x^{t-1}e^{-x}dx, \Re(t)>0$

$\int$: *Line integral*

$x \equiv y[p]$: *x is congruent to y* mod *p*

$n!$: *for* $n \in \mathbb{N}, n! = n.(n-1)....2.1$

$h \sim g$: $\lim_{x\to\infty} \frac{h(x)}{g(x)} = 1$

$\mu(n)$:*The Mobius function*

H :*The Hadamard matrix*

$L_i(t)$:*The logarithme integral of t*, $L_i(t) = \int_2^t \frac{dx}{\log x}$

$h(x) = o(g(x))$: $\lim_{x\to\infty} \frac{h(x)}{g(x)} = 0$

$h(x) = O(g(x))$: $|h(x)| \leq C|g(x)|$ *for some C and all values of* $x \succ x_0$ *for some* x_0

References

[1] R. J. Anderson, On the Mertens conjecture for cusp forms, Mathematika 26 (1979),

[2] R. J. Anderson, A note on the Mo. .bius sum function, in preparation.

[3] R. J. Anderson and H. M. Stark, Oscillation theorems in: M. I. Knopp (ed.), Analytic Number Theory, Proc. Conf. Temple Univ., Lecture Notes in Mathematics, Springer, 1981.

[4] P. T. Bateman, J. W. Brown, R. S. Hall, K. E. Kloss and R. M. Stemmler, Linear relations connecting the imaginary parts of the zeros of the zeta function, pp. 11-19 in: A. O. L. Atkin and B. J. Birch (editors), Computers in number theory, Proc. Atlas Symp., Acad. Press, New York, 1971.

[5] R. P. Boas, Jr., and M. Kac, Inequalities for Fourier transforms of positive functions, Duke Math. J. 12 (1945),

[6] H. Bohr and B. Jessen, One more proof of Kronecker's Theorem, J. London Math. Soc., 7 (1932)

[7] R. P. Brent, A Fortran multiple-precision arithmetic package, ACM Trans. Math. Software 4 (1978),

[8] H. Cohen, Arithmétique et informatique, Astérisque, 61 (1979), 57-61.

[9] H. G. Diamond, Two oscillation theorems, pp. 113-118 in: A. A. Gioia and D. L. Goldsmith (editors), The theory of arithmetic functions, Lecture Notes in Mathematics, Springer, 1972.

[10] H. M. Edwards, Riemann's Zeta Function, Academic Press, 1974.

[11] L. J. Goldstein, A necessary and sufficient condition for the Riemann hypothesis for zeta functions attached to eigenfunctions of the Hecke operator, Acta Arith. 15 (1968/69),

[12] H. M. Edwards, 2001, Riemann's Zeta Function, Dover Publications, Inc.

[13] G. H. Hardy and E. M. Wright, 1979, An Introduction To Theory Of Numbers, Oxford, England: Clarendon Press

[14] B. B. Mandelbrot, 1977, The Fractal Geometry Of Nature, W. H. Freeman

[15] B. Riemann, 1859, On The Number Of Prime Numbers Less Than A Given Quantity, Berlin Academy of Sciences

[16] Journal of Number Theory Volume 129, Issue 10, October 2009, Pages 2557-2568 Journal of Number Theory - Distribution of powers of the partition function modulo ℓj

[17] La méthode mathématique de Dedekind. [Article] sem-linkMohamed Sinaceur Revue d'histoire des sciences Année 1979

[18] Les constructions des nombres réels, Jacqueline Boniface, IREM - Histoire des mathématiques, Éd. Ellipses, Paris - 2002

[19] Traité de MATHÉMATIQUES spéciales - Tome 2, analyse, G. Cagnac, E. Ramis, J. Commeau, Éd. Masson & Cie - Paris, 1972

[20] '' Was sind und was sollen die Zahlen" une analyse du traité par David E. Joyce :

aleph0.clarku.edu

[21] La création des nombres, par Hourya Benis Sinaceur, philosophe (CNRS).

[22] Les constructions des nombres réels dans le mouvement d'arithmétisation de l'analyse par Jacqueline Boniface, Univ. Toulouse 2.

[23] '' ERNEST COUMET '' Les diagrammes de Venn Mathématiques et sciences humaines, tome 10 (1965),

[24] Venn, John. 1880. "On the Diagrammatic and Mechanical Representation of Propositions and Reasonings," in Dublin Philosophical Magazine and Journal of Science. 1–18. vol. 9, 59, 1880.

[25] Cambridge University Library. 1975. Catalogue of a collection of books on logic: presented to the library by John Venn. Link title. New York : B. Franklin.

[26] Collection of manuscripts, Gonville and Caius College Library, Cambridge.

[27] Krúger, L. (1987). The slow rise of probabilism..., in Krúger et. al. (eds.), The Probabilistic Revolution, Vol. 1, MIT Press, Cambridge, Mass.,

[28] A Heeffer, The origin of the problems in Euler's Algebra, Bull. Belg. Math. Soc. Simon Stevin 13 (5) (2006),

[29] LA CONJECTURE DE GOLDBACH POUR LES NOMBRES IMPAIRS Le 20 mai 2013 - Ecrit par Bruno Duchesne

[30] Yvette Smith, « Katherine Johnson : The Girl Who Loved to Count » [archive], sur nasa.gov — Her calculations proved as critical to the success of the Apollo Moon landing program and the

start of the Space Shuttle program, as they did to those first steps on the country's journey into space. "Katherine G. Johnson Biography" [archive]. Biography.com. 10 octobre 2016.

[31] WILLIAM OUGHTRED A GREAT SEVENTEENTH-CENTURY TEACHER OF MATHEMATICS BY FLORIAN CAJORI, Ph.D. Professor of Mathematics Colorado College CHICAGO LONDON THE OPEN COURT PUBLISHING COMPANY 1916

[32] LES MATHÉMATIQUES ARABES DU 8E AU 15E SIÈCLE, Adolf P. Youschkevitch,Éd. Vrin – CNRS – Paris 1976

[33] L'algèbre arabe : Genèse d'un art, par Ahmed Djebbar, Éd. Vuibert/Adapt, Paris – 2005 L'algèbre d'Al-Khawarizmi et les méthodes indienne et grecque, Léon Rodet, 1878

[34] JOURNAL ARTICLE Eléments d'analyse de Karl Weierstrass Pierre Dugac, H. A. Schwarz, Ihe Weierstrass and S. Lie Archive for History of Exact Sciences Vol. 10, No. 1/2 (28.VI.1973),

[35] Anne-Françoise Jaccottet, « Hypatie d'Alexandrie entre réalité historique et récupérations idéologiques : réflexions sur la place de l'Antiquité dans l'imaginaire moderne », Études de lettres, 1-2 | 2010, 139-158.

[36] Lamirande, Emilien, « Hypatie, Synésios et la fin des dieux. L'histoire et la fiction », Studies in Religion / Sciences religieuses, 18 (1989), p. 467-489.

[37] Rougé, Jean, « La politique de Cyrille d'Alexandrie et le meurtre d'Hypatie », Cristianesimo nella storia, 11 (1990),

[38] ENCYCLOPÆDIA BRITANNICA : britannica.com/science/mathematics

[39] [Rediffusion d'un article publié en 2012] LA CONJECTURE DE HADAMARD (I) Piste bleue Le 19 avril 2020 - Ecrit par Shalom Eliahou